NUREG-1437
Supplement 38, Vol. 4

I0494102

Generic Environmental Impact Statement for License Renewal of Nuclear Plants

Supplement 38

Regarding Indian Point Nuclear Generating Units Nos. 2 and 3

Final Report Supplemental Report and Comment Responses

Office of Nuclear Reactor Regulation

AVAILABILITY OF REFERENCE MATERIALS
IN NRC PUBLICATIONS

U.S.NRC

United States Nuclear Regulatory Commission

Protecting People and the Environment

NUREG-1437
Supplement 38, Vol. 4

Generic Environmental Impact Statement for License Renewal of Nuclear Plants

Supplement 38

Regarding Indian Point Nuclear Generating Units Nos. 2 and 3

Final Report Supplemental Report and Comment Responses

Manuscript Completed: May 2013
Date Published: June 2013

Office of Nuclear Reactor Regulation

ABSTRACT

This supplement to the final supplemental environmental impact statement (FSEIS) for the proposed license renewal of Indian Point Nuclear Generating Unit Nos. 2 and 3 incorporates new information that the U.S. Nuclear Regulatory Commission (NRC) staff has obtained since the publication of the FSEIS in December 2010.

This supplement includes corrections to impingement and entrainment data presented in the FSEIS, revised conclusions regarding thermal impacts based on newly available thermal plume studies, and an update of the status of the NRC's consultation under Section 7 of the Endangered Species Act with the National Marine Fisheries Service regarding the shortnose sturgeon (*Acipenser brevirostrum*) and Atlantic sturgeon (*Acipenser oxyrinchus oxyrinchus*).

TABLE OF CONTENTS

LIST OF FIGURES

LIST OF TABLES

EXECUTIVE SUMMARY

BACKGROUND

By letter dated April 23, 2007, Entergy Nuclear Operations, Inc. (Entergy) submitted an application to the U.S. Nuclear Regulatory Commission (NRC) to issue renewed operating licenses for Indian Point Nuclear Generating Unit Nos. 2 and 3 (IP2 and IP3) for additional 20-year periods.

Under Title 10 of the Code of Federal Regulations (10 CFR) 51.20(b)(2) and the National Environmental Policy Act of 1969, as amended (NEPA), the renewal of a power reactor operating license requires preparation of an environmental impact statement (EIS) or a supplement to an existing EIS. In addition, 10 CFR 51.95(c) states that the NRC shall prepare an EIS, which is a supplement to the Commission's NUREG–1437, "Generic Environmental Impact Statement for License Renewal of Nuclear Plants," issued May 1996.

The NRC published its final supplemental environmental impact statement (FSEIS) for IP2 and IP3 in December 2010. After the NRC published the FSEIS, the staff identified new information that necessitated changes to its assessments in the FSEIS. This new information is derived from the following:

- Entergy provided comments on the FSEIS that included new information on the entrainment and impingement field data units of measure.

- Entergy provided comments on the Essential Fish Habitat Assessment that also included new information on the data units of measure.

- Entergy completed and submitted to the New York State Department of Environmental Conservation a new study that characterizes the IP2 and IP3 thermal plume.

To address this new information, the NRC staff has prepared this supplement to the FSEIS in accordance with 10 CFR 51.92(a)(2) and (c), which address preparation of a supplement to a final EIS for proposed actions that have not been taken, under the following conditions:

- There are new and significant circumstances or information relevant to environmental concerns and bearing on the proposed action or its impacts, or

- The NRC staff determines, in its opinion, that preparation of a supplement will further the purposes of NEPA.

In addition to supplementing the FSEIS for the reasons stated above, the NRC is also taking this opportunity to document the completion of the consultation process under Section 7 of the Endangered Species Act of 1973, as amended (ESA), with the National Marine Fisheries Service (NMFS) regarding the shortnose sturgeon (*Acipenser brevirostrum*) and the Atlantic sturgeon (*Acipenser oxyrinchus oxyrinchus*) population in the New York Bight.

PROPOSED ACTION

The proposed action remains the same as that stated in the FSEIS (at pages 1–6 and 1–7):

> The proposed Federal action is renewal of the operating licenses for IP2 and
> IP3 (IP1 was shut down in 1974). IP2 and IP3 are located on approximately
> 239 acres of land on the east bank of the Hudson River at Indian Point,
> Village of Buchanan, in upper Westchester County, New York, approximately

24 miles north of the New York City boundary line. The facility has two Westinghouse pressurized–water reactors. IP2 is currently licensed to generate 3216 megawatts thermal (MW(t)) (core power) with a design net electrical capacity of 1078 megawatts electric (MW(e)). IP3 is currently licensed to generate 3216 MW(t) (core power) with a design net electrical capacity of about 1080 MW(e). IP2 and IP3 cooling is provided by water from the Hudson River to various heat loads in both the primary and secondary portions of the plants. The current operating license for IP2 expires on September 28, 2013, and the current operating license for IP3 expires on December 12, 2015. By letter dated April 23, 2007, Entergy submitted an application to the NRC (Entergy 2007a) to renew the IP2 and IP3 operating licenses for an additional 20 years.

PURPOSE AND NEED FOR ACTION

The purpose and need for action remains the same as stated in the FSEIS (at page 1–7):

> Although a licensee must have a renewed license to operate a reactor beyond the term of the existing operating license, the possession of that license is just one of a number of conditions that must be met for the licensee to continue plant operation during the term of the renewed license. Once an operating license is renewed, State regulatory agencies and the owners of the plant will ultimately decide whether the plant will continue to operate based on factors such as the need for power or matters within the State's jurisdiction—including acceptability of water withdrawal, consistency with State water quality standards, and consistency with State coastal zone management plans—or the purview of the owners, such as whether continued operation makes economic sense.
>
> Thus, for license renewal reviews, the NRC has adopted the following definition of purpose and need (GEIS Section 1.3):
>
>> The purpose and need for the proposed action (renewal of an operating license) is to provide an option that allows for power generation capability beyond the term of a current nuclear power plant operating license to meet future system generating needs, as such needs may be determined by State, utility, and where authorized, Federal (other than NRC) decision makers.
>
> This definition of purpose and need reflects the Commission's recognition that, unless there are findings in the safety review required by the Atomic Energy Act of 1954, as amended, or findings in the NEPA environmental analysis that would lead the NRC to reject a license renewal application, the NRC does not have a role in the energy–planning decisions of State regulators and utility officials as to whether a particular nuclear power plant should continue to operate. From the perspective of the licensee and the State regulatory authority, the purpose of renewing the operating licenses is to maintain the availability of the nuclear plant to meet system energy requirements beyond the current term of the plant's licenses.

ABBREVIATIONS, ACRONYMS, AND SYMBOLS

°C	degree(s) Celsius
°F	degree(s) Fahrenheit
ADAMS	Agencywide Documents Access and Management System
BSS	Beach Seine Survey
CFR	Code of Federal Regulations
CHGEC	Central Hudson Gas and Electric Corporation
CMR	conditional mortality rate
DPS	distinct population segment
EIS	environmental impact statement
EMR	entrainment mortality rate
Entergy	Entergy Nuclear Operations, Inc.
ESA	Endangered Species Act of 1973, as amended
FSEIS	final supplemental environmental impact statement
FSS	Fall Shoals Survey
ft	feet
GEIS	NUREG-1437, "Generic Environmental Impact Statement for License Renewal of Nuclear Plants"
IMR	impingement mortality rate
IP2 and IP3	Indian Point Nuclear Generating Unit Nos. 2 and 3
IPEC	Indian Point Energy Center
ITS	incidental take statement
LRS	Long River Survey
m	meter(s)
NEPA	National Environmental Policy Act of 1969
NMFS	National Marine Fisheries Service
NRC	U.S. Nuclear Regulatory Commission
NYB	New York Bight
NYCRR	New York Codes, Rules, and Regulations
NYSDEC	New York State Department of Environmental Conservation
RIS	representative important species
SEIS	supplemental environmental impact statement
SOC	strength of connection
SPDES	State Pollutant Discharge Elimination System

WOE	weight of evidence
YOY	young–of–year

1.0 INTRODUCTION

The U.S. Nuclear Regulatory Commission (NRC) staff prepared this supplement to the final supplemental environmental impact statement (FSEIS) for Indian Point Nuclear Generating Units 2 and 3 (IP2 and IP3) in accordance with Title 10 of the *Code of Federal Regulations* (10 CFR) 51.92(a)(2) and (c), which address the preparation of a supplement to an FSEIS for proposed actions that have not been taken, if the following conditions apply:

- There are new and significant circumstances or information relevant to environmental concerns and bearing on the proposed action or its impacts, or

- The NRC staff determines, in its opinion, that preparation of a supplement will further the purposes of NEPA.

The NRC staff prepared this supplement to the FSEIS because it received new data, analyses, and comments from several sources that potentially changed, and in some cases did change, the staff's conclusions in the FSEIS. This supplement contains the text, tables, and figures that changed as the result of this new information.

Three sources provided information that changed the staff's conclusions in the FSEIS.

First, in comments to the NRC dated March 29, 2011, Entergy Nuclear Operations, Inc. (Entergy) (Entergy 2011b, AKRF 2011b) provided new information regarding the entrainment and impingement field data that it had previously provided to the NRC for its aquatic resource impact assessment in Entergy (2007), a December 2007 supplement to its license renewal application. In its letter dated March 29, 2011, Entergy (2011b) said that these changes would:

> ...not alter, but rather confirm, NRC's ultimate conclusion in the FSEIS that potential impacts to aquatic species as a result of theoretical entrainment and impingement at IPEC are no more than MODERATE.

Second, comments submitted on behalf of Entergy (Goodwin Proctor 2011, AKRF 2011a) on the FSEIS and the NRC staff's Essential Fish Habitat Assessment contained related new information. When the NRC staff considered this information, the staff found that the information necessitated some minor changes to the aquatic ecology findings in Sections 4.1.2 through 4.1.3 of the FSEIS and Appendices H and I. Chapter 2 of this supplement provides corrected tables and conclusions resulting from the NRC staff's analysis of the new information. Where specific changes or corrections to FSEIS information occur, this supplement references the affected FSEIS section, page, and line numbers.

Third, since the publication of the FSEIS, Entergy submitted to the New York State Department of Environmental Conservation (NYSDEC) a triaxial plume study (Swanson et al. 2011a) as part of its State Pollutant Discharge Elimination System (SPDES) permit renewal application. Entergy undertook this study in response to the NYSDEC's 2010 Notice of Denial (NYSDEC 2010). Based on this new information, as well as Entergy's response to the NYSDEC staff's comments on the study (Mendelsohn et al. 2011, Swanson et al. 2011b) and the NYSDEC staff's conclusions regarding its review of the study and response to comments (NYSDEC 2011), the NRC staff has revised its conclusions regarding the impacts of heat shock to aquatic species. Chapter 3 of this supplement presents these revised conclusions.

In addition to supplementing the FSEIS for the reasons stated above, the staff is also taking this opportunity to update the status of consultations under Section 7 of the Endangered Species Act of 1973, as amended (ESA) with the National Marine Fisheries Service (NMFS). Chapter 4 of this supplement updates the information contained in Section 4.6.1 of the FSEIS to document the completion of consultation regarding the shortnose sturgeon (*Acipenser brevirostrum*) and Atlantic sturgeon (*Acipenser oxyrinchus oxyrinchus*) in the New York Bight (NYB), and summarizes the biological opinion and associated incidental take statement (ITS) (NMFS 2011e) that NMFS issued in January 2013 as a result of that consultation.

The NRC staff issued a draft supplement to the FSEIS on June 26, 2012, which was made available for public comment for 45 days. Based on comments received, the NRC staff amended the draft supplement to the FSEIS, as necessary, and published this final supplement to the FSEIS. The comments received, and the NRC staff's responses to those comments, are presented in Appendix A of this supplement.

Where appropriate, bold text indicates specific text corrections or additions to the FSEIS and bold strikeout indicates deletions from the text. Change bars (vertical lines in the page margin) indicate changes that were made to the text of the draft supplement to the FSEIS, prior to issuing this final supplement.

2.0 IMPINGEMENT AND ENTRAINMENT DATA CORRECTIONS

2.1 Corrections to Section 4.1.2, "Entrainment of Fish and Shellfish in Early Lifestages," and Its Related Appendices

In a letter to the NRC dated March 29, 2011 (Entergy 2011b; AKRF 2011b), Entergy provided new information supplementing the entrainment and impingement field data that it had previously provided to the NRC for its aquatic resource impact assessment. This new information appears in "Technical Review of FSEIS for Indian Point Nuclear Generating Unit Nos. 2 and 3" (AKRF 2011b). In its technical review, AKRF (2011b) stated that the units of the entrainment catch densities provided by Entergy are expressed as the number caught per 1,000 cubic meters (m^3). Because Entergy did not originally provide the units used in the FSEIS to assess impacts, the NRC staff believed the units to be the number caught per m^3 based on historical documents provided by Entergy, comments by Entergy and its consultants on the draft SEIS, and phone conversations among Entergy, Entergy's consultants, and the NRC staff. Thus, the entrainment losses the FSEIS reported for each of the representative important species (RIS) used in the NRC staff's analysis are too large by a factor of 1,000.

In the FSEIS, the NRC staff estimated the number entrained for a given week as the product of the mean density entrained and the combined weekly flow for IP2 and IP3. The error in the entrainment catch density directly affects Figure 4–3 in Section 4.1.2, and the error is repeated in Figure H–5 in Appendix H. In these figures, the total number entrained on the right axis should be in units of numbers × 10^8 instead of numbers × 10^{11}. The corrected Figures 4–3 and H–5 appear below. In addition, these changes affect two portions of text in the FSEIS.

Lines 2 and 3 of page 4-14 in the FSEIS are corrected as follows:

> The total number of identified fish entrained has decreased at a rate of 187 ~~billion~~ million fish per year since 1984.

Lines 1–3 of page H-22 in the FSEIS are corrected as follows:

> Linear regression (n=6; p<0.01) indicated that the number of identified fish entrained decreased at a rate of 187 ~~billion~~ million fish per year, a result consistent with the decrease observed in the number of fish impinged.

The change in units of the entrainment catch densities also affects the 75th percentile of the number of each life stage entrained and the annual estimate of the number entrained presented in Tables I–39 and I–42 of Appendix I. In Table I–39, the units should be numbers × 10^3 instead of numbers × 10^6. In Table I–42, the units should be numbers in the thousands instead of numbers in the millions. The corrected tables appear below.

Figure 4–3 on page 4-15 in the FSEIS is corrected as follows:

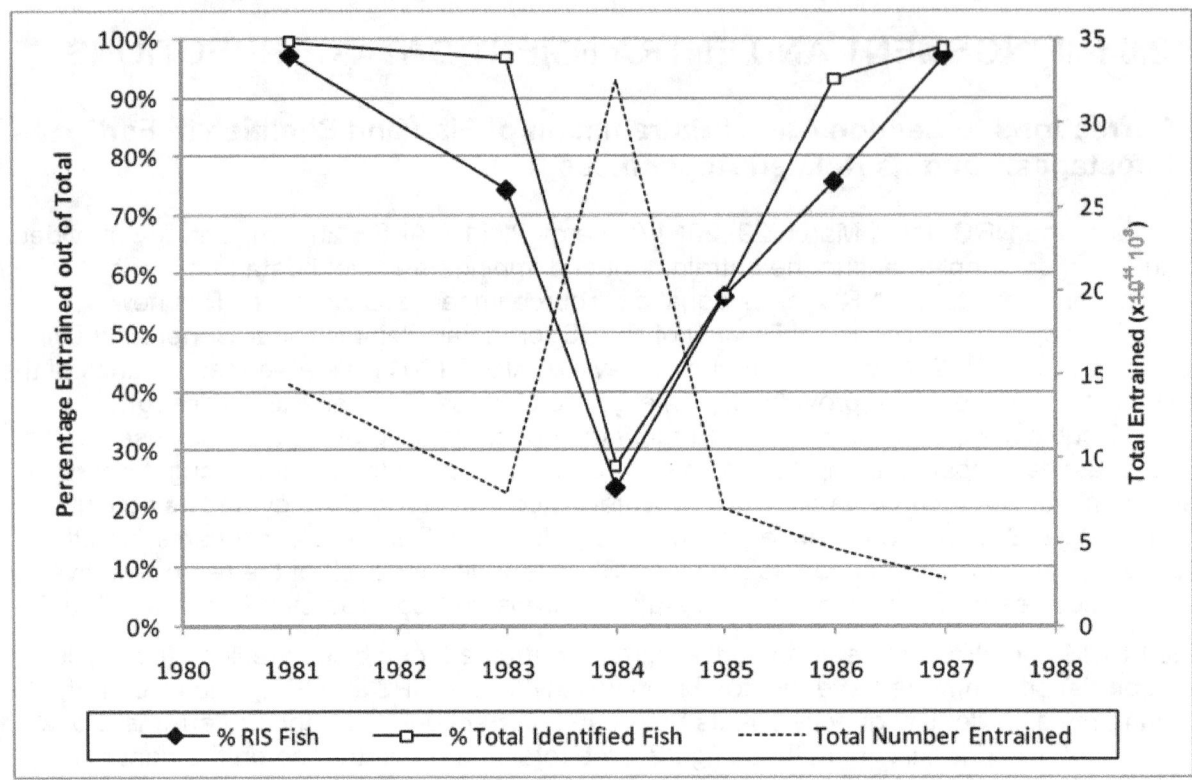

Figure 4–3. Percentage of entrainment composed of RIS fish and total identified fish relative to the estimated total entrainment at IP2 and IP3 combined (data from Entergy 2007b)

Figure H–5 on page H-23 in the FSEIS is identical to Figure 4–3 in the FSEIS and is corrected as follows:

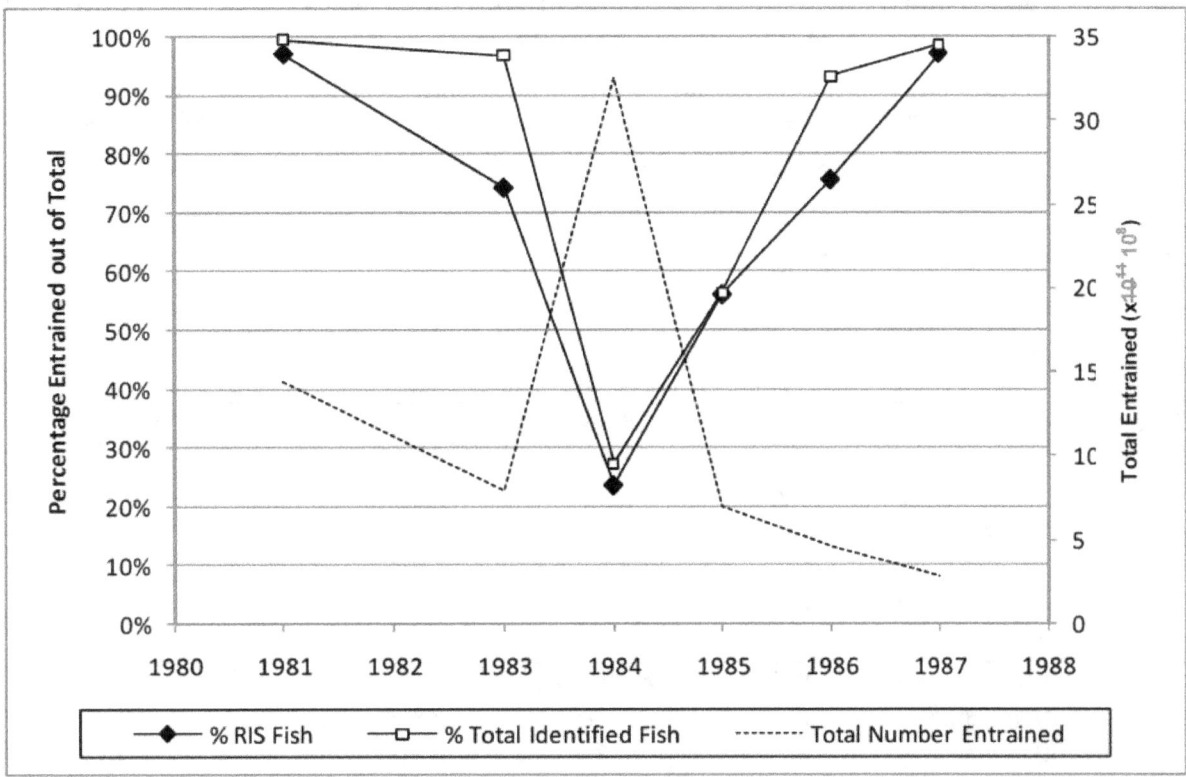

Figure H–5. Percentage of entrainment composed of RIS fish and total identified fish relative to the estimated total entrainment at IP2 and IP3 combined (data from Entergy 2007b)

Impingement and Entrainment Corrections

Table I–39 on page I-54 in the FSEIS is corrected as follows:

Table I-39. Percentage of Each Life Stage Entrained by Season and the Contribution of Major Taxa Represented in the Samples.

Calculations are based on the 75th percentile over years (1981 and 1983–1987) of each season's number of fish entrained. No entrainment sampling occurred in October–December.

Life Stage	Season 1 Jan–Mar	Season 2 Apr–Jun	Season 3 Jul–Sep	75th Percentile over Years
EGG	3%	20%	78%	210,801× ~~10⁶~~ 10^3
Rainbow Smelt	99%	2%	0%	
Bay Anchovy	0%	92%	100%	
White Perch	0%	4%	<1%	
Alosa species	1%	2%	0%	
YOLK–SAC LARVA	8%	89%	3%	23,140× ~~10⁶~~ 10^3
Atlantic Tomcod	100%	0%	0%	
Herring Family	0%	91%	<1%	
Bay Anchovy	0%	2%	94%	
Striped Bass	0%	5%	1%	
Hogchoker	0%	0%	3%	
POST YOLK–SAC LARVA	<1%	52%	48%	618,393× ~~10⁶~~ 10^3
Atlantic Tomcod	100%	<1%	0%	
Alosa species	0%	37%	<1%	
Bay Anchovy	0%	11%	58%	
Anchovy Family	0%	2%	39%	
White Perch	0%	12%	1%	
Striped Bass	0%	17%	1%	
Herring Family	0%	20%	<1%	
JUVENILE	2%	44%	54%	10,989× ~~10⁶~~ 10^3
White Perch	96%	10%	10%	
Atlantic Tomcod	0%	67%	2%	
Weakfish	0%	1%	50%	
Bay Anchovy	0%	1%	17%	
Rainbow Smelt	0%	9%	3%	
Striped Bass	0%	6%	5%	
Anchovy Family	0%	1%	4%	
Alosa species	0%	2%	2%	
White Catfish	4%	<1%	0%	
Blueback Herring	0%	<1%	3%	
UNDETERMINED STAGE	10%	77%	13%	4,469× ~~10⁶~~ 10^3
Atlantic Tomcod	100%	<1%	0%	
Morone species	0%	88%	2%	
Bay Anchovy	0%	9%	83%	
Anchovy Family	0%	0%	10%	
Alosa species	0%	0%	4%	

The title of Table I–42 on page I-58 of the FSEIS is corrected as follows:

**Table I–42   Annual Estimated Number of RIS Entrained at IP2 and IP3
(~~millions~~ thousands of fish)**

The contents of the table remain accurate and, therefore, are not duplicated in this supplement.

2.2 Corrections to Section 4.1.3, "Combined Effects of Impingement and Entrainment," and Its Related Appendices

In a letter to the NRC dated March 29, 2011, Entergy (2011b) provided new information (in AKRF 2011b) regarding the units associated with the catch density data from the Long River Survey (LRS) and the Fall Shoals Survey (FSS) that Entergy (2007) had previously submitted to the NRC for its aquatic resource impact assessment. In AKRF's (2011b) technical review, the units of the catch densities are expressed as the number caught per 1,000 m^3. Entergy did not provide the units for these densities when it originally submitted the data to the NRC. The NRC staff based the units it used in the FSEIS to assess impacts (i.e., number caught per m^3) on information in the mathematical construction of these measures provided in Central Hudson Gas and Electric Corporation (CHGEC) et al. (1999). Thus, the NRC staff overestimated the annual standing crop from the LRS and FSS in the FSEIS for each of the representative important species (RIS) by a factor of 1,000. The NRC staff then used the estimates of the annual standing crop and the estimated entrainment losses to estimate a conditional entrainment mortality rate (EMR), a parameter in the models used in the strength-of-connection (SOC) analysis.

The NRC staff described the calculation of the standing crop from the LRS and FSS in Appendix I, Section I.2.2, of the FSEIS. The NRC staff estimated the LRS and FSS weekly standing crop as the weekly density of fish caught multiplied by the IP2 and IP3 region river volume. The error in the density units for the LRS and FSS produced incorrect estimates of the combined standing crop used in the denominator of the estimated EMR in the FSEIS. The NRC staff also used entrainment losses as input to the numerator and the denominator of the EMR estimates. Because both the numerator and the denominator of the estimated EMR were too large by a factor of 1,000, only those estimates for two RIS (spottail shiner (*Notropis hudsonius*) and white catfish (*Ameiurus catus*)), in which the Beach Seine Survey (BSS) contributed more of the standing crop, were seriously affected. For the remaining RIS, to which the BSS contributed little, the errors in units largely cancelled because of the construction of the EMR as a ratio of the number entrained (which was 1,000 times too large) to the number at risk (number in the river plus the number entrained, both of which were 1,000 times too large). The amount and direction of change in the EMR depends on the relative contributions from the three sampling programs—BSS, FSS, and LRS.

The NRC staff used the EMR in its assessment of the SOC and, ultimately, to determine the final weight-of-evidence (WOE) assessment of the combined effects of impingement and entrainment from IP2 and IP3. The unit of measure error affects the staff's conclusion of High SOC for spottail shiner, but not the conclusion of Low SOC for white catfish. The NRC staff reran the SOC Monte Carlo simulations using the corrected EMRs, and, based on the corrected data, now finds a Low SOC for the spottail shiner. Further, based on the WOE assessment of the combined effects of impingement and entrainment from IP2 and IP3, the NRC staff concludes that the impacts of impingement and entrainment on the spottail shiner are SMALL rather than LARGE.

The changes to the SOC analysis affect FSEIS Table 4–4 (presented below) and several lines of text in Section 4.1.3.3. However, Section 4.1.3.5 is not affected by these changes.

Impingement and Entrainment Corrections

Lines 41–43 on page 4-20 of the FSEIS are corrected as follows:

Based on the WOE assessment (Table 4–4), the NRC staff concludes that impacts to American shad, Atlantic menhaden, Atlantic sturgeon, Atlantic tomcod, bay anchovy, bluefish, gizzard shad, shortnose sturgeon, spottail shiner, striped bass, white catfish, and blue crab are SMALL.

Lines 1–3 on page 4-21 of the FSEIS are corrected as follows:

The NRC staff concludes that impacts to alewife, rainbow smelt, and weakfish are MODERATE. The staff concludes that impacts to blueback herring, hogchoker, spottail shiner, and white perch are LARGE.

Lines 30–41 on page 4-21 of the FSEIS are removed as follows:

Spottail Shiner

The NRC staff concludes that a Large impact is present for YOY spottail shiner because a detectible population decline occurred in the river wide (1 of 3) and river segment (1 of 1) data sets, and the strength of connection with the IP2 and IP3 cooling system is high. The habitat for the spottail shiner includes small streams, lakes, and large rivers, including the Hudson. This species feeds primarily on aquatic insect larvae, zooplankton, benthic invertebrates, and fish eggs and larvae, and is the prey of striped bass. Spottail shiners spawn from May to June or July (typically later for the northern populations) over sandy bottoms and stream mouths (Smith 1985; Marcy et al. 2005); water chestnut (Trapa natans) beds provide important spawning habitat (CHGEC 1999). Individuals older than 3 years are rare, although some individuals may live 4 or 5 years (Marcy et al. 2005). Spottail shiner is not a marine or anadromous species, so coastal population trend data are not available.

Table 4–4 on page 4-23 of the FSEIS is corrected as follows:

Table 4–4. Impingement and Entrainment Impact Summary for Hudson River YOY RIS

Species	Population Trend Line of Evidence	Strength of Connection Line of Evidence	Impacts of IP2 and IP3 Cooling Systems on YOY RIS
Alewife	Variable	High	Moderate
American Shad	Detected Decline	Low	Small
Atlantic Menhaden	Unresolved[a]	Low[b]	Small
Atlantic Sturgeon	Unresolved[a]	Low[b]	Small
Atlantic Tomcod	Detected Decline	Low	Small
Bay Anchovy	Undetected Decline	High	Small
Blueback Herring	Detected Decline	High	Large
Bluefish	Detected Decline	Low	Small
Gizzard Shad	Unresolved[a]	Low[b]	Small
Hogchoker	Detected Decline	High	Large
Rainbow Smelt	Variable	High	Moderate–Large[c]
Shortnose Sturgeon	Unresolved[a]	Low[b]	Small
Spottail Shiner	Detected Decline	~~High~~ Low	~~Large~~ Small
Striped Bass	Undetected Decline	High	Small
Weakfish	Variable	High	Moderate
White Catfish	Variable	Low	Small
White Perch	Detected Decline	High	Large
Blue Crab	Unresolved[a]	Low[b]	Small

(a) Population trend could not be established because of a lack of river survey data.

(b) Monte Carlo simulation could not be conducted because of the low rate of entrainment and impingement; a Low Strength of connection was concluded.

(c) Section 4.1.3.3 provides supplemental information.

Because of the new information regarding the units of the data for entrainment density and the density of fish caught during the LRS and FSS, the NRC staff corrected the estimates of EMR for American shad (*Alosa sapidissima*), bay anchovy (*Anchoa mitchilli*), hogchoker (*Trinectes maculates*), white catfish, and white perch (*Morone americana*) reported in Appendices H and I. The staff's conclusions of the SOC for these RIS, however, remain unchanged. These changes affect several lines of text in Sections H.1.3.2 and H.1.3.3 and Tables H–16 and H–17, as described below.

Lines 11–12 on page H-47 in Section H.1.3.2 of the FSEIS are corrected as follows:

The results of this analysis indicated a High strength of connection for ~~nine~~ eight species (Table H–16).

Lines 15–16 on page H-47 in Section H.1.3.2 of the FSEIS are corrected as follows:

For ~~four~~ five RIS, the strength of connection was Low (minimal evidence of connection).

Lines 5–10 on page H-49 in Section H.1.3.3 of the FSEIS are corrected as follows:

Based on the WOE assessment (Table H–17), the NRC staff concludes that the impact levels are Small for ~~eleven~~ 12 species: American shad, Atlantic menhaden, Atlantic sturgeon, Atlantic tomcod, bay anchovy, bluefish, gizzard shad, shortnose sturgeon, spottail shiner, striped bass, white catfish, and blue crab. Further, the staff concludes that the impacts are Moderate for three species: alewife, rainbow smelt, and weakfish. Finally, the staff concludes that the impacts are Large for ~~four~~ three species: blueback herring, hogchoker, ~~spottail shiner;~~ and white perch.

Lines 26–38 on page H-50 in Section H.1.3.3 of the FSEIS are removed as follows:

Spottail Shiner

~~The NRC staff concludes that a Large impact is present for YOY spottail shiner because a detectible population decline occurred in the river wide (1 of 3) and river segment (1 of 1) data sets, and there was a high strength of connection with the IP2 and IP3 cooling system. The habitat for the spottail shiner includes small streams, lakes, and large rivers, including the Hudson. This species feeds primarily on aquatic insect larvae, zooplankton, benthic invertebrates, and fish eggs and larvae, and is the prey of striped bass. Spottail shiners spawn from May to June or July (typically later for the northern populations) over sandy bottoms and stream mouths (Smith 1985; Marcy et al. 2005); water chestnut (Trapa natans) beds provide important spawning habitat (CHGEC 1999). Individuals older than 3 years are rare, but there is evidence of individuals living four or five years (Marcy et al. 2005). Coastal population trend data were not available for this species.~~

Table H–16 on page H-48 of the FSEIS is corrected as follows:

Table H–16. Weight of Evidence for the Strength–of–Connection Line of Evidence for YOY RIS Based on the Monte Carlo Simulation

RIS	Strength of Connection	RIS	Strength of Connection
Alewife	High	Hogchoker	High
American Shad	Low	Rainbow Smelt	High
Atlantic Menhaden	Cannot be Modeled[a]	Shortnose Sturgeon	Cannot be Modeled[a]
Atlantic Sturgeon	Cannot be Modeled[a]	Spottail Shiner	~~High~~ Low
Atlantic Tomcod	Low	Striped Bass	High
Bay Anchovy	High	Weakfish	High
Blueback Herring	High	White Catfish	Low
Bluefish	Low	White Perch	High
Gizzard Shad	Cannot be Modeled[a]	Blue Crab	Cannot be Modeled[a]

[a] Estimates for model parameters were unavailable or information was lacking. Strength of connection assumed to be Low based on review of impingement and entrainment data.

Table H–17 on page H-49 of the FSEIS is corrected as follows:

Table H–17. Impingement and Entrainment Impact Summary for Hudson River YOY RIS

Species	Population Trend Line of Evidence	Strength of Connection Line of Evidence	Impacts of IP2 and IP3 Cooling Systems on YOY RIS
Alewife	Variable	High	Moderate
American Shad	Detected Decline	Low	Small
Atlantic Menhaden	Unresolved[a]	Low[b]	Small
Atlantic Sturgeon	Unresolved[a]	Low[b]	Small
Atlantic Tomcod	Detected Decline	Low	Small
Bay Anchovy	Undetected Decline	High	Small
Blueback Herring	Detected Decline	High	Large
Bluefish	Detected Decline	Low	Small
Gizzard Shad	Unresolved[a]	Low[b]	Small
Hogchoker	Detected Decline	High	Large
Rainbow Smelt	Variable	High	Moderate–Large[c]
Shortnose Sturgeon	Unresolved[a]	Low[b]	Small
Spottail Shiner	Detected Decline	~~High~~ Low	~~Large~~ Small
Striped Bass	Undetected Decline	High	Small
Weakfish	Variable	High	Moderate
White Catfish	Variable	Low	Small
White Perch	Detected Decline	High	Large
Blue Crab	Unresolved[a]	Low[b]	Small

[a] Population Line of Evidence could not be established using WOE; therefore, population Line of Evidence could range from small to large.

[b] Strength of connection could not be established using Monte Carlo simulation; therefore, strength of connection was based on the rate of entrainment and impingement.

[c] Section 4.1.3.3 provides supplemental information.

In addition to Tables I–39 and I–42, presented earlier, the new information about the units of measure affects tables in Appendix I. The corrected Table I–40, Table I–41, Table I–43, Table I–46, and Table I–47 in Appendix I of the FSEIS appear on the following pages.

Table I–40 on page I-56 of the FSEIS is corrected as follows:

Table I–40.   Method for Estimating Taxon-Specific Entrainment Mortality Rate (EMR) Based on River Segment 4 Standing Crop for the Strength of Connection Analysis

Property of Method		Number Entrained	River Segment 4 Standing Crop
Input Data	Variables	Mean density organisms entrained by IP2 and IP3 (# per 1000 m^3) Volume of cooling water withdrawn by IP2 and IP3 (1000 m^3/min)	LRS density (by life stage) FSS density of YOY (# per 1000 m^3) BSS density of YOY (# per haul) River Segment 4 volume (m^3) River Segment 4 shorezone surface area (m^2)
	Frequency	Per week of sampling	Per week of sampling
Summary Statistics	Seasonal (Year specific)	Sum of weekly estimates of number of organisms entrained by IP2 and IP3	Sum of weekly standing crop estimates
	Annual	Sum of Season 1, 1986, with each year's totals from Season 2 and Season 3	Sum of seasonal standing crop estimates for River Segment 4
	EMR	$$\frac{\text{75th Percentile Annual Number Entrained}}{\text{75th Percentile (Annual Number Entrained + Annual Standing Crop)}}$$	
	Units of numerator and denominator of EMR	# of organisms	
Years of Data		1981 and 1983–1987	1981 and 1983–1987
Life Stages		Eggs, Larvae, and Juveniles	Eggs, Larvae, and Juveniles (YOY)
Taxonomic Substitutions		Alewife, blueback herring, and unidentified alosids treated collectively as river herring	
		Unidentified anchovy spp. (species, plural) allocated to bay anchovy	
		Unidentified *Morone* spp. allocated proportionally to striped bass and white perch	

The title of **Table I–41 on page I-57** of the FSEIS is corrected as follows:

Table I–41.   Estimated Annual Standing Crop of Eggs, Larvae, and Juvenile RIS Within River Segment 4 (~~millions~~thousands of fish)

The contents of the table remain accurate and, therefore, are not duplicated in this supplement.

Table I–43 on page I-59 of the FSEIS is corrected as follows:

Table I–43. Estimate of the River Segment 4 Entrainment Mortality Rate (EMR) and the 95 Percent Confidence Limits for the Riverwide Entrainment CMR (1974–1997)

Taxa	75th Percentile Annual Number Entrained (number x ~~10⁹~~ 10⁶)	75th Percentile of Number at Risk (number x ~~10⁹~~ 10⁶)	EMR	Riverwide CMR for Entrainment at IP2 and IP3	
				Lower 95% Confidence Limit	Upper 95% Confidence Limit
Alewife and Blueback Herring	94.9	1003	0.095	0.00747	0.0324
American Shad	0.357	~~8.43~~ 9.26	~~0.042~~ 0.039	0	0.016696
Atlantic Menhaden	0	NA	NA	Not Modeled	
Atlantic Sturgeon	0	NA	NA	Not Modeled	
Atlantic Tomcod	7.65	210	0.036	0.152	0.234
Bay Anchovy	439	~~2064~~ 2065	~~0.213~~ 0.212	0.0925	0.140
Bluefish	0.00291	1.13	0.003	Not Modeled	
Gizzard Shad	0	NA	NA	Not Modeled	
Hogchoker	1.87	~~4.83~~ 4.84	~~0.386~~ 0.385	Not Modeled	
Rainbow Smelt	7.07	27.4	0.258	Not Modeled	
Shortnose Sturgeon	0	NA	NA	Not Modeled	
Spottail Shiner	0.00295	~~0.00838~~ 0.0937	~~0.352~~ 0.031	0.0802	0.104
Striped Bass	71.4	676	0.106	0.181	0.276
Weakfish	3.90	7.17	0.544	Not Modeled	
White Catfish	0.00965	~~0.0848~~ 0.0388	~~0.114~~ 0.249	Not Modeled	
White Perch	63.5	~~840~~ 841	~~0.076~~ 0.075	0.0568	0.108

Table I–46 on page I-61 of the FSEIS is corrected as follows:

Table I–46. Parameter Values Used in the Monte Carlo Simulation

RIS	Survey Used	Linear Slope (*r*)	Upper 95% Confidence Limit of the Slope	Error Mean Square from Regression	CV of Density Data (1979-1990)	EMR	IMR
Alewife	BSS	-0.030	-0.014	0.570	1.245	0.095	0.0020
American Shad	BSS	-0.069	-0.059	0.350	0.744	~~0.042~~ 0.39	0.0005
Atlantic Tomcod	FSS	-0.040	-0.026	0.490	1.035	0.036	0.0300
Bay Anchovy	FSS	-0.075	-0.061	0.505	0.598	~~0.213~~ 0.212	0.0040
Blueback Herring	BSS	-0.024	-0.009	0.530	1.488	0.095	0.0040
Bluefish	BSS	-0.038	-0.022	0.580	0.692	0.003	0.0005
Hogchoker	FSS	-0.034	-0.018	0.580	1.679	~~0.386~~ 0.385	0.0005
Rainbow Smelt	FSS	0.012	0.041	0.576	1.452	0.258	0.0005
Spottail Shiner	BSS	-0.017	-0.005	0.430	1.293	~~0.352~~ 0.031	0.0070
Striped Bass	BSS	0.040	0.052	0.420	0.528	0.106	0.0080
Weakfish	FSS	-0.047	-0.031	0.560	1.085	0.544	0.0005
White Catfish	FSS	0.007	0.010	0.100	3.520	~~0.114~~ 0.249	0.0005
White Perch	BSS	-0.062	-0.045	0.610	0.848	~~0.076~~ 0.075	0.0320

Table I–47 on page I-63 of the FSEIS is corrected as follows:

Table I–47. Quartiles of the Relative Difference in Cumulative Abundance and Conclusions for the Strength-of-Connection from the Monte Carlo Simulation

Taxa	Number of Years	$N_0 = 1000$			$N_0 = 1 \times 10^8$			Strength of Connection Conclusion
		Median	Q1	Q3	Median	Q1	Q3	
Alewife	20	0.33	0.11	0.59	0.32	0.06	0.55	High
	27	0.36	0.15	0.56	0.33	0.14	0.53	
American Shad	20	~~0.07~~ 0.08	~~-0.04~~ -0.03	~~0.18~~ 0.20	~~0.09~~ 0.08	~~-0.02~~ -0.03	~~0.20~~ 0.19	Low
	27	~~0.08~~ 0.07	-0.01	~~0.16~~ 0.15	~~0.08~~ 0.07	~~0.00~~ -0.01	0.16	
Atlantic Tomcod	20	0.14	-0.04	0.32	0.17	-0.01	0.38	Low
	27	0.18	0.04	0.32	0.18	0.02	0.33	
Bay Anchovy	20	~~0.21~~ 0.19	~~0.09~~ 0.08	~~0.32~~ 0.31	0.20	0.08	0.31	High
	27	~~0.18~~ 0.19	0.10	~~0.26~~ 0.28	0.18	~~0.10~~ 0.09	~~0.27~~ 0.28	
Blueback Herring	20	0.30	0.02	0.60	0.28	0.02	0.60	High
	27	0.43	0.16	0.67	0.40	0.14	0.64	
Bluefish	20	0.13	-0.04	0.29	0.14	-0.03	0.30	Low
	27	0.14	0.02	0.29	0.16	0.01	0.30	
Hogchoker	20	~~0.71~~ 0.72	~~0.39~~ 0.37	~~1.05~~ 1.06	~~0.74~~ 0.76	~~0.41~~ 0.42	~~1.10~~ 1.09	High
	27	~~0.81~~ 0.76	~~0.53~~ 0.50	~~1.10~~ 1.09	~~0.77~~ 0.84	~~0.46~~ 0.56	~~1.06~~ 1.13	
Rainbow Smelt	20	0.77	0.33	1.25	0.81	0.35	1.34	High
	27	0.93	0.52	1.38	1.03	0.63	1.46	
Spottail Shiner	20	~~0.59~~ 0.20	~~0.33~~ -0.07	~~0.88~~ 0.43	~~0.58~~ 0.18	~~0.23~~ -0.06	~~0.90~~ 0.42	~~High~~ Low
	27	~~0.64~~ 0.22	~~0.36~~ 0.01	~~0.88~~ 0.42	~~0.62~~ 0.23	~~0.35~~ 0.01	~~0.87~~ 0.46	
Striped Bass	20	0.45	0.09	0.76	0.45	0.12	0.78	High
	27	0.62	0.27	1.02	0.66	0.31	1.01	
Weakfish	20	0.62	0.39	0.87	0.66	0.42	0.90	High
	27	0.63	0.43	0.84	0.64	0.43	0.83	
White Catfish	20	~~0.19~~ 0.40	~~-0.36~~ -0.20	~~0.76~~ 0.98	~~0.05~~ 0.37	~~-0.46~~ -0.18	~~0.66~~ 1.00	Low
	27	~~0.09~~ 0.39	~~-0.41~~ -0.15	~~0.58~~ 0.91	~~0.09~~ 0.37	~~-0.43~~ -0.19	~~0.58~~ 0.99	
White Perch	20	~~0.16~~ 0.18	~~0.01~~ 0.03	~~0.32~~ 0.35	~~0.20~~ 0.19	~~0.04~~ 0.03	~~0.35~~ 0.34	High
	27	~~0.18~~ 0.19	~~0.06~~ 0.07	~~0.31~~ 0.30	~~0.20~~ 0.17	~~0.07~~ 0.06	~~0.31~~ 0.30	

3.0 ASSESSMENT OF THERMAL IMPACTS

In the FSEIS, the NRC staff concluded that the potential impacts of the cooling water discharge from IP2 and IP3 on aquatic species could range from SMALL to LARGE because the staff did not have enough information to quantify the extent and magnitude of the IP2 and IP3 thermal plume. Since publication of the FSEIS, the NRC has obtained additional information from Entergy regarding the thermal plume that enables the staff to make a more informed conclusion regarding thermal impacts.

In January 2011, Entergy submitted to the NYSDEC a preliminary report on a triaxial plume study (Swanson et al 2011a) as part of its SPDES permit renewal application. Entergy undertook this study in response to the NYSDEC's 2010 Notice of Denial (NYSDEC 2010), which noted that Entergy's previous thermal study (Swanson et al. 2010) did not directly address the period of highest river temperatures, and as such, would require additional confirmatory monitoring to determine whether any modeled results accurately show compliance with thermal standards. The NYSDEC provided Entergy with comments on the new Swanson et al. (2011a) study in March 2011. Within the same month, Mendelsohn et al. (2011) and Swanson et al. (2011b) prepared responses to the NYSDEC staff's review of the study. In a letter dated May 16, 2011, NYSDEC (2011) notified NYSDEC Judges M.E. Villa and D.P. O'Connell that it had finished reviewing the data and information contained in both the study and the response to NYSDEC's comments and that, based on this information and applicable regulations, the NYSDEC staff had determined the following:

> …a thermal mixing zone in the Hudson River near Indian Point not to exceed a maximum of seventy-five (75) acres in total size during any time of a given year (6 NYCRR §704.3) will provide reasonable assurance of compliance with water quality standards and criteria for thermal discharges set forth in 6 NYCRR §§704.1 and 704.2, respectively.

Based on Swanson et al.'s (2011a) triaxial thermal plume study, Mendelsohn et al.'s (2011) and Swanson et al.'s (2011b) responses to NYSDEC staff comments on the study, and NYSDEC staff's (2011) conclusions regarding the study, the NRC staff has revised its discussion of and conclusions regarding thermal impacts to aquatic species, which appear in Section 4.1.4 of the FSEIS.

Lines 16–26 on page 4-30 in Section 4.1.4.3 of the FSEIS are changed as follows:

> ~~Entergy has been engaged in discussions with the NYSDEC concerning the thermal impacts of IP2 and IP3 cooling water system operation. As a result of those discussions, the NRC staff notes that Entergy recently performed a triaxial thermal study of the Hudson River from September 9 to November 1 of 2009 (Entergy 2010). Given the months involved in this study, the study period did not include days with the highest average annual water temperature. Entergy has indicated that it will perform modeling of the river based on its field data in order to determine whether the power plant is in compliance with conditions of its permit; it also indicated that it may conduct additional monitoring in 2010. The NYSDEC, in its recent Notice of Denial of Water Quality Certification, indicated that additional verification of any modeled results would be necessary (NYSDEC 2010). Entergy did conduct additional studies in 2010. This issue continues to be subject to NYSDEC authority and review.~~

In February 2010, Entergy submitted to NYSDEC a preliminary report (Swanson et al. 2010) on a triaxial thermal study of the Hudson River performed during the period of September 9 to November 1, 2009. Because the study did not directly address the period of highest river temperatures, the NYSDEC directed Entergy to perform additional confirmatory monitoring to determine whether any modeled results accurately show compliance with thermal standards (NYSDEC 2010). In January 2011, Entergy submitted to the NYSDEC a new triaxial plume study (Swanson et al. 2011a).

In the new study, Swanson et al. (2011a) reported that the extent and shape of the thermal plume varied greatly, primarily in response to tidal currents. For example, the plume (illustrated as a 4°F (2.2°C) temperature increase or ΔT isotherm in Figure 5–6 of Swanson et al. 2011a) generally followed the eastern shore of the Hudson River and extended northward from IP2 and IP3 during flood tide and southward from IP2 and IP3 during ebb tide. Depending on tides, the plume can be reasonably easily identified and can reach a portion of the near-shore bottom or be largely confined to the surface of the river.

Temperature measurements reported by Swanson et al. (2011a) generally show that the warmest water in the thermal plume is close to the surface, and plume temperatures tend to decrease with depth. A cross-river survey conducted in front of IP2 and IP3 captured one such incident during spring tide on July 13, 2010 (Figure 3–28 in Swanson et al. 2011a). Across most of the river, water temperatures were close to 82°F (28°C), often with warmer temperatures near the surface and cooler temperatures near the bottom. The IP2 and IP3 thermal plume at that point was clearly defined and extended about 1,000 feet (ft) (300 meters (m)) from shore on a cross-river transect of about 3800 ft (1150 m) (interpreted from the figure). Surface water temperatures in the plume reached about 85°F (29°C). Maximum river depth along the measured transect is approximately 50 ft (15 m).

A temperature contour plot at a cross-river transect at IP2 and IP3 illustrates a similar condition on July 11, 2010, during slack before flood tide (Figure 1–10 in Swanson et al. 2011b). Here, the thermal plume is evident to about 2,000 ft (600 m) from the eastern shore (the location of the IP2 and IP3 discharge) and extends to a depth of about 35 ft (11 m) along the eastern shore. The river here is more than 4,500 ft (1,400 m) wide. Bottom temperatures above 82°F (28°C) were confined to about the first 250 ft (76 m) from shore. In that small area, bottom water temperatures might also exceed 86°F (30°C); elsewhere, bottom water temperatures were about 80°F (27°C). The NRC staff notes, however, that these limited-area conditions would not last long, as they would change with the tidal cycle.

In response to NYSDEC's review of the IP2 and IP3 thermal studies (Swanson et al. 2011a), Mendelsohn et al. (2011) modeled the maximum area and width of the thermal plume (defined by the 4°F (2.2°C) ΔT isotherms) in the Hudson River. Mendelsohn et al. (2011) reported that for four cross-river transects near IP2 and IP3, the maximum cross-river area of the plume would not exceed 12.3 percent of the river cross-section, and the maximum cross-river width of the plume would not exceed 28.6 percent of the river width (Table 3–1 in Mendelsohn et al. 2011).

Swanson et al. (2011a) concluded that IP2 and IP3 are in compliance with NYSDEC water quality standards set forth at 6 NYCRR Part 704.

After line 43 on page 4-31 of Section 4.1.4.4 of the FSEIS, the following text is to be added:

In response to the NYSDEC's 2010 Notice of Denial (NYSDEC 2010), Entergy submitted a new triaxial plume study (Swanson et al. 2011a) to the NYSDEC in January 2011. NYSDEC provided Entergy with comments on the new study (Swanson et al. 2011a) in March 2011. Within the same month, Mendelsohn et al. (2011) and Swanson et al. (2011b) prepared responses to the NYSDEC staff's review of the study. In a May 2011 letter (NYSDEC 2011), NYSDEC staff notified NYSDEC Judges M.E. Villa and D.P. O'Connell that NYSDEC staff had finished reviewing the data and information contained in both the study and the response to NYSDEC's comments and that, based on this information and applicable regulations, NYSDEC staff had determined the following:

a thermal mixing zone in the Hudson River near Indian Point not to exceed a maximum of seventy-five (75) acres in total size during any time of a given year (6 NYCRR §704.3) will provide reasonable assurance of compliance with water quality standards and criteria for thermal discharges set forth in 6 NYCRR §704.1 and 704.2, respectively.

Lines 2–26 on page 4-32 in Section 4.1.4.5 of the FSEIS are corrected as follows:

~~In the absence of a completed thermal study proposed by NYSDEC (or an alternative proposed by Entergy and accepted by NYSDEC), existing information must be used to determine the appropriate thermal impact level to sensitive life stages of important aquatic species. Since NYSDEC modeling in the FEIS (NYSDEC 2003a) indicates that discharges from IP2 and IP3 could raise water temperatures to a level greater than that permitted by water quality criteria that are a component of existing NYSDEC permits, the staff must conclude that adverse impacts are possible. Cold water fish species such as Atlantic tomcod and rainbow smelt may be particularly vulnerable to temperature changes caused by thermal discharges. The population of both species has declined, and rainbow smelt may have been extirpated from the Hudson River. The NYSDEC's issuance of a SPDES permit provides a basis to conclude that the thermal impacts of IP2 and IP3 discharges could meet applicable regulatory temperature criteria. The NYSDEC's recent pronouncements and its ongoing re-examination of this issue create uncertainty, and this issue is currently being addressed in NYSDEC administrative proceedings. Accordingly, in the absence of specific studies, and in the absence of results sufficient to make a determination of a specific level of impact, the NRC staff concludes that thermal impacts from IP2 and IP3 potentially could range from SMALL to LARGE depending on the extent and magnitude of the~~

~~thermal plume, the sensitivity of various aquatic species and life stages likely to encounter the thermal plume, and the probability of an encounter occurring that could result in lethal or sublethal effects. This range of impact levels expresses the uncertainty accruing from the current lack of studies and data. Either additional thermal studies or modeling and verification of Entergy's 2009 thermal study might generate data to further refine or modify this impact level. For the purposes of this Final SEIS, the NRC staff concludes that the impact level could range from SMALL to LARGE. This conclusion is meant to satisfy NRC's NEPA obligations and is not intended to prejudice any determination the NYSDEC may reach in response to new studies and information submitted to it by Entergy.~~

NRC regulations for license renewal environmental reviews establish the primary role of the U.S. Environmental Protection Agency (EPA) (or States, when applicable) in water quality regulations as they relate to impacts on aquatic species. As such, the assessment of impacts from heat shock is within the purview of the responsible government agency. In the case of IP2 and IP3, NYSDEC is the responsible agency.

NYSDEC regulations at 6 NYCRR Part 704 establish specific standards that apply to thermal discharges within the State of New York. The standards are set to "assure the protection and propagation of a balanced, indigenous population of shellfish, fish, and wildlife in and on the body of water" to which heated water is discharged (6 NYCRR 704.1(a)). Section 4.1.4.4 of this FSEIS supplement describes the thermal plume studies (Swanson et al. 2010, 2011a) that Entergy submitted to NYSDEC and NYSDEC's (2011) conclusions regarding these studies. NYSDEC concluded that the results of the thermal plume studies provide reasonable assurance that the IP2 and IP3 discharge is in compliance with NYSDEC's water quality standards and criteria for thermal discharges.

Based on Entergy's thermal plume studies and NYSDEC's conclusions, the NRC staff concludes that the impacts from heat shock to aquatic resources of the lower Hudson River would be SMALL.

This change in the NRC staff's conclusion regarding thermal impacts (heat shock) also affects the Abstract, Executive Summary, Alternatives, and Summary sections of the FSEIS. The NRC staff has revised parts of these sections, as described below.

Line 37 on page iii through line 2 on page iv of the FSEIS Abstract are changed as follows:

Overall effects from entrainment and impingement are likely to be MODERATE, and impacts from heat shock are likely to be SMALL. ~~Impacts from heat shock potentially range from SMALL to LARGE depending on the conclusions of thermal studies proposed by the New York State Department of Environmental Conservation (NYSDEC).~~

Lines 33–39 on page xviii of the FSEIS Executive Summary are changed as follows:

The NRC staff concludes that the potential environmental effects for most of these issues are of SMALL significance in the context of the standards set forth in the GEIS with ~~three~~ two exceptions—entrainment~~,~~ and impingement~~, and heat shock from the facility's heated discharge~~. The NRC staff jointly assessed the impacts of entrainment and impingement to be MODERATE based on NRC's analysis of representative important species. ~~Impacts from heat shock potentially range from SMALL to LARGE depending on the conclusions of thermal studies conducted by Entergy and submitted to the NYSDEC.~~

Line 43 on page 8-8 through line 3 on page 8-9 of Section 8.1.1.2 are changed as follows:

Because the closed-cycle cooling system discharges a smaller volume of water, and because the water is cooler than in a once-through system, the extent of thermal impacts ~~which could range from SMALL to LARGE for the current once-through system, given uncertainty in the facility's thermal impacts~~ would remain SMALL ~~be reduced~~. ~~Thus, the effects of thermal shock also decline.~~

Lines 35–40 on page 9-4 of Section 9.1 are changed as follows:

The NRC staff concludes that the potential environmental effects for ~~9~~ 10 of the 12 categorized issues are of SMALL significance in the context of the standards set forth in the GEIS. The NRC staff concludes that the combined impacts from impingement and entrainment (each a separate issue) are MODERATE. ~~Impacts from heat shock could range from SMALL to LARGE, based on the large uncertainties discussed in Chapter 4.~~

Lines 8–13 on page 9-5 of Section 9.1 are changed as follows:

For issues of MODERATE ~~or LARGE~~ significance (i.e., issues related to aquatic ecology), mitigation measures are addressed both in Chapter 4 and in Chapter 8 as alternatives based on determinations in the draft New York State Department of Environmental Conservation (NYSDEC) State Pollutant Discharge Elimination System (SPDES) permit proceeding, Clean Water Act Section 401 proceeding, and in draft policy statements published by the State.

4.0 SECTION 7 CONSULTATION

At the time the NRC staff published the FSEIS, the NRC and NMFS had not completed Section 7 consultation under the Endangered Species Act of 1973, as amended (ESA) for the shortnose sturgeon (*Acipenser brevirostrum*). During the course of the Section 7 consultation, the NRC staff obtained more studies and information on the thermal plume (previously discussed in Chapter 3 of this document). As a result, the NRC staff has revised its conclusions regarding thermal impacts to the shortnose sturgeon based on this new thermal modeling information. Section 2.2.5.5 of the FSEIS, which includes the shortnose sturgeon's life history, remains unchanged. The staff identified one correction to Section 4.6.1 of the FSEIS, shown below.

In addition to supplementing the FSEIS for the reasons stated in Chapter 1 of this supplement, the staff is also taking this opportunity to provide an update on the status of its consultation with NMFS related to Indian Point Nuclear Generating Unit Nos. 2 and 3 (IP2 and IP3). This chapter provides an update on the Section 7 consultation history provided in Section 4.6.1 of the FSEIS, as well as a summary of the biological opinion that NMFS issued in October 2011 as a result of consultation. This chapter also provides a summary of the reinitiation of consultation regarding the Atlantic sturgeon (*Acipenser oxyrinchus oxyrinchus*). Consultation with NMFS regarding the Atlantic sturgeon was reinitiated as a result of NMFS's February 2012 listing of Atlantic sturgeon as an endangered species under the ESA and concluded in January 2013 with NMFS's issuance of a final biological opinion for both the shortnose and Atlantic sturgeon, which included an Incidental Take Statement (ITS).

4.1 Corrections to Section 4.6.1, "Aquatic Special Status Species"

In the FSEIS, the NRC staff concluded that the potential impacts of heated discharge from IP2 and IP3 on shortnose sturgeon could not be determined because the staff did not have enough information to quantify the extent and magnitude of the IP2 and IP3 thermal plume. Since publication of the FSEIS, the NRC staff has obtained additional information on the IP2 and IP3 thermal plume. Chapter 3 of this document describes the new thermal plume information. Based on Swanson et al.'s (2011a) triaxial thermal plume study, Mendelsohn et al.'s (2011) and Swanson et al.'s (2011b) responses to NYSDEC staff comments on the study, and NYSDEC staff's (2011) conclusions regarding the study, the NRC staff has revised its discussion regarding thermal impacts to shortnose sturgeon, which appears in Section 4.6.1 of the FSEIS.

Lines 40–43 on page 4-58 in Section 4.6.1 of the FSEIS are changed as follows:

> ~~The potential impacts of thermal discharges on shortnose and Atlantic sturgeon cannot determined at this time because additional studies are required to quantify the extent and magnitude of the thermal plume, as discussed in Section 4.1.4 of this SEIS.~~
>
> In July 2011, the NRC (2011c) supplemented its analysis of the thermal effects from IP2 and IP3 on the shortnose sturgeon that was presented in NRC's (2010) December 2010 revised biological assessment. The NRC staff's (2011c) supplement to the revised biological assessment considered newly available thermal plume information (Swanson et al. 2011a, 2011b; Mendelsohn et al. 2011; NYSDEC 2011) as well as various studies on shortnose sturgeon biology and thermal preferences (Dadswell 1979; Dadswell et al. 1984; Heidt and Gilbert 1978; Ziegeweid et al. 2008a, 2008b). In its July 2011 supplement, the NRC (2011c)

concluded that the proposed license renewal of IP2 and IP3 is not likely to adversely affect the Hudson River population of shortnose sturgeon.

NMFS issued its biological opinion in October 2011 (NMFS 2011e). In its biological opinion, NMFS concluded that shortnose sturgeon are likely to avoid the small area of water elevated above the species' preferred temperature range and that—

> it is extremely unlikely that these minor changes in behavior will preclude shortnose sturgeon from completing any essential behaviors such as resting, foraging or migrating or that the fitness of any individuals will be affected.

Based on the NRC's (2011c) previous analysis and NMFS's (2011e) biological opinion, the NRC staff concludes that the heated discharge resulting from the proposed IP2 and IP3 license renewal would have SMALL impacts on the shortnose sturgeon.

Lines 13–20 on page 4-59 and Lines 1–16 on page 4-60 in Section 4.6.1 of the FSEIS are modified as follows:

The NRC staff reviewed information from the site audit, Entergy's ER for the IP2 and IP3 site, other reports, and information from NMFS. Based on the WOE information presented in Table 4-4, The NRC staff concludes that the impacts associated with the IP2 and IP3 cooling system are Small for both Atlantic and shortnose sturgeon. The population trend LOE evaluation was unresolved because the Hudson River monitoring programs were not designed to catch either species. The NRC staff was also unable to determine the strength of connection for either species using the Monte Carlo simulation modeling. Because historical impingements of sturgeon have been relatively low, especially for shortnose sturgeon, the NRC staff concluded

that the strength of connection was low. Based on the WOE analysis described above, a determination of Moderate or Large impact is not supported, and the NRC staff concludes that the impacts of an additional 20 years (beyond the current term) of operation and maintenance of the site on aquatic species that are Federally listed as threatened or endangered is SMALL. The NRC staff is sending a revised biological assessment (BA) of the impacts of license renewal on the shortnose sturgeon to NMFS to review as this SEIS goes to press (the BA will be publicly available at ML102990042). Should NMFS determine that continued operation of IP2 and IP3 has the potential to adversely impact the shortnose sturgeon, NMFS will issue a biological opinion. Included in the biological opinion would be any reasonable and prudent measures that the applicant could undertake, as well as the terms and conditions for the applicant to comply with the formal Section 7 consultation. Possible mitigation measures could range from a resumption of monitoring to determine the number of shortnose sturgeon impinged at IP2 and IP3 to changes in the cooling water intake system, as described in Section 4.1.5 of this FEIS. Additionally, as described in Chapter 8, the installation of cooling towers could reduce impingement, entrainment, and thermal impacts for all aquatic resources, including those that are Federally listed.

In addition to the WOE information provided in Table 4–4, the staff examined the new information from the ESA Section 7 consultations with NMFS to determine the level of impact for the purposes of NEPA. Because NMFS (2013) finds that license renewal would not change the status or trend of the Hudson River population of shortnose sturgeon or the species as a whole, the NRC staff finds that the level of impact would be SMALL for this species. For Atlantic sturgeon, NMFS finds that license renewal would not change the status or trend of the Hudson River population of Atlantic sturgeon or the status and trend of the NYB DPS as a whole. NMFS (2013) calculates that the highest observed annual impingement of Atlantic sturgeon at the traveling screens would represent about 0.5 percent of the Hudson River origin juveniles. This potential reduction would not be observable or noticeable through any population study. Therefore, the staff finds that the level of impact would be SMALL for Atlantic sturgeon. Furthermore, development and implementation of an appropriate monitoring plan for these species at IP2 and IP3 would help ensure protection of these species. Based on the NRC's (2011C) previous analysis, as corrected herein, and NMFS's (2013) biological opinion, the staff finds that the level of impact for aquatic special status species would be SMALL.

4.2 History of Section 7 Consultation for Shortnose Sturgeon

Under Section 7 of the ESA, the NRC staff (2008b) initiated consultation with NMFS in a letter dated December 22, 2008, upon publication of the draft supplemental environmental impact statement (SEIS) and the staff's (NRC 2008a) original biological assessment, which found that the relicensing of IP2 and IP3 could adversely affect the shortnose sturgeon, which had been listed as endangered under the ESA in 1967. In response to that biological assessment, on February 24, 2009, NMFS (2009) requested additional information from the NRC. NMFS stated that it required this information before it could begin formal consultation. On July 1, 2009, the NRC staff obtained the relevant information from Entergy (2009). On August 10, 2009, the NRC (2009) provided that information (including revised impingement data) to NMFS and stated that the data would be addressed in the FSEIS and in a revised biological assessment. The NRC staff published its FSEIS in December 2010 and transmitted its revised biological assessment to NMFS on December 10, 2010 (NRC 2010b).

On February 16, 2011, NMFS (2011) formally responded to the NRC staff's letter of December 10, 2010, and stated that (1) NMFS currently has all the information it needs to complete a formal consultation, (2) NMFS considers formal consultation to have begun on December 16, 2010, (3) NMFS expects the consultation will conclude within 90 days after it began (i.e., by March 16, 2011) unless extended, and (4) NMFS expects to issue its biological opinion by April 30, 2011. On March 1, 2011, Entergy (2011a) formally notified the NRC staff that it will participate in the consultation process and requested a 45-day extension of the consultation conclusion date in accordance with 50 CFR 402.14(e).

In teleconferences on March 9 and March 11, 2011, NMFS and the NRC staff discussed extending the consultation to allow time for Entergy to submit additional information on the shortnose sturgeon pertinent to the consultation (NRC 2011h). NMFS formally extended the consultation period in a March 16, 2011, letter (NMFS 2011a) for a period of 60 days until June 29, 2011, in accordance with 50 CFR 402.14(e). On April 18, 2011, the NRC staff (2011a) held a Category 1 public meeting during which Entergy presented a data synthesis on the shortnose sturgeon updated with the most recent annual Hudson River monitoring reports. On April 28, 2011, Entergy (2011c) formally submitted to the NRC the information it had presented during this public meeting.

On June 16, 2011, the NRC staff learned that Entergy had submitted a final, verified triaxial thermal model to NYSDEC concerning aquatic conditions at IP2 and IP3. The staff also learned that NYSDEC had relied on that model and Entergy's associated information to reach conclusions about thermal conditions at Indian Point for inclusion in a draft SPDES permit (NYSDEC 2011). The NRC staff (2011b) brought this information to NMFS's attention in an e-mail to NMFS on June 16, 2011.

The NRC staff held three teleconferences with NMFS and Entergy during the weeks of June 20 and June 27, 2011 (NRC 2011d). On June 20, 2011, the NRC staff and NMFS discussed the NRC's statutory authority to implement terms and conditions or reasonable and prudent measures identified in a biological opinion. On June 22, 2011, the NRC staff, NMFS, and Entergy discussed NMFS's outstanding questions on thermal impacts, impingement, and entrainment of prey species and the design of the IP2 and IP3 cooling system. The NRC staff also requested that Entergy formally submit to the NRC the thermal modeling information that Entergy had given to NYSDEC. By letter dated June 29, 2011, Entergy (2011d) formally submitted to the NRC various documents related to the thermal studies it had conducted. During a teleconference on June 29, 2011, the NRC staff, NMFS, and Entergy addressed questions that had arisen during the teleconference on June 22, 2011, and the parties agreed to a revised consultation schedule in which the consultation would end by September 20, 2011, provided that Entergy and the NRC staff would supply NMFS with the information related to NMFS's outstanding questions in a timely manner. The NRC staff (2011c) supplemented its revised biological assessment on July 26, 2011, as a result of the information that Entergy submitted to the staff on June 29, 2011.

NMFS (2011b) issued a draft biological opinion on August 26, 2011. In an e-mail dated September 6, 2011, the NRC staff provided NMFS with Entergy's comments on the draft biological opinion (NRC 2011f). In a separate e-mail on the same day, the staff submitted its comments on the draft biological opinion (NRC 2011e). The NRC staff stated that its comments on the draft biological opinion were complete and that it would respond to the procedural issues raised in NMFS's cover letter to the draft biological opinion in a separate letter. On September 19, 2011, NMFS (2011c) requested more time to complete the final biological opinion. On September 20, 2011, the NRC staff (2011g) sent its letter addressing the issues NMFS had raised in the cover letter to its draft biological opinion.

NMFS (2011d, 2011e) issued its final biological opinion for shortnose sturgeon on October 14, 2011 (referred to as the 2011 biological opinion), which concluded the Section 7 consultation for the IP2 and IP3 license renewal. The NMFS 2011 biological opinion is discussed below.

4.3 Summary of the National Marine Fisheries Service's Biological Opinion for Shortnose Sturgeon

NMFS's 2011 biological opinion (2011d, 2011e) included an ~~incidental take statement~~ITS for shortnose sturgeon and stipulated a number of reasonable and prudent measures, as well as terms and conditions with which the NRC and Entergy must comply to be exempt from prohibitions of Section 9 of the ESA.

Under the 2011 biological opinion, IP2 and IP3 may take up to the following numbers of shortnose sturgeon during the terms of their renewed operating licenses, which NMFS assumed would not begin before the completion of the initial operating licenses for IP2 and IP3:

- 6 shortnose sturgeon at Unit 1
- 104 shortnose sturgeon at Unit 2
- 58 shortnose sturgeon at Unit 3

NMFS included Unit 1, even though it is not in operation, because Unit 2 uses water from the Unit 1 intake as service water.

The 2011 biological opinion stipulated four reasonable and prudent measures that require Entergy to (1) implement an NMFS-approved monitoring program, (2) release all live sturgeon back to the Hudson River, (3) transfer any dead sturgeon to NMFS for necropsy, and (4) report all shortnose sturgeon impingements or sightings to NMFS. The terms and conditions provided the NRC and Entergy with more specific details on how the reasonable and prudent measures must be carried out. The terms and conditions can be found on pages 64–67 of the biological opinion. If the NRC renews the IP2 and/or IP3 licenses, compliance with the terms and conditions of the biological opinion (as later revised) will be required, as appropriate[1].

4.4 Reinitiation of Consultation Due to NMFS's Listing of Atlantic Sturgeon

On February 6, 2012, NMFS listed five distinct population segments (DPSs) of the Atlantic sturgeon (*Acipenser oxyrinchus oxyrinchus*) under the ESA (77 FR 5880; 77 FR 5914). In the Hudson River near Indian Point, Atlantic sturgeon primarily belong to the New York Bight DPS, which NMFS listed as endangered. The NRC staff had previously addressed the environmental impacts of license renewal on the Atlantic sturgeon in the final SEIS and had requested that NMFS conduct a Section 7 conference with the staff regarding the Atlantic sturgeon, which was proposed for listing at that time. On May 16, 2012, in response to the listing, the NRC staff (2012) prepared and submitted a biological assessment to NMFS, along with a request to reinitiate Section 7 consultation for the newly-listed Atlantic sturgeon. The NRC staff expects to continue consultation with NMFS in 2012 regarding Atlantic sturgeon at IP2 and IP3, and will consider the results of that consultation, as appropriate.

The NRC provided much of the information needed for this reinitiated consultation in its FSEIS (NRC 2010a) and the revised biological assessment for shortnose sturgeon (NRC 2010b) and its supplement (NRC 2011). Entergy (2011e) and its consultants (Barnthouse et al. 2011) provided additional information to NMFS on shortnose and Atlantic sturgeon in the Hudson River, the characteristics of IP2 and IP3, and the facility's effects on the two sturgeon species. Entergy (2012) also provided lists and reviews of reports providing information on the effects of IP2 and IP3 on Atlantic sturgeon.

In its May 16, 2012, biological assessment, the NRC (2012a) concluded that

> ...operation of IP2 and IP3 may affect, but is not likely to adversely affect, the Atlantic sturgeon during the remainder of the current operating license period and the 20-year license renewal term (through September 28, 2033 and December 12, 2035, respectively), if license renewal is approved.

NMFS considers reinitiation of consultation to have begun on May 17, 2012, the day after it received the NRC staff's biological assessment. On July 3, 2012, in a telephone call between NMFS and the NRC staff, the NRC staff clarified that its was requesting reinitiated consultation to consider the effects to shortnose sturgeon and the five DPS of Atlantic sturgeon due to operation of IP2 and IP3 during both the remainder of the present license terms and the possible renewed license terms. On July 23, 2012, Entergy supplied additional information on Atlantic and shortnose sturgeon impingement at IP2 and IP3 (AKRF et al. 2012). The NRC staff and NMFS, by mutual agreement, then

---

[1] The 2011 biological opinion stated: "This [incidental take statement] ITS applies to the extended operating period, beginning at the date that the facility begins to operate under the terms of a new license and extending through the expiration date of that license." (NMFS 2011e)

extended the consultation to allow time to review and incorporate the new information in accordance with 50 CFR 402.14(e). NMFS transmitted the draft biological opinion to the NRC for review on October 26, 2012, and the NRC staff then transmitted it to Entergy. On November 9, 2012, the NRC (2012b) transmitted to NMFS both Entergy's and the NRC staff's comments on the draft biological opinion. The NRC staff requested, via a conference call, that the consultation period be extended for 7 days on November 26, 2012. On December 5, 2012, NMFS requested that the consultation be extended to January 9, 2013, to allow time for the NRC and NMFS to discuss language in the ITS. During a conference call on January 8, 2013, the NRC and Entergy provided additional comments related to the ITS, and Entergy submitted additional comments on wording to NMFS on January 9, 2013. On January 9, 2013, the NRC staff and Entergy requested an extension of consultation until January 30, 2013, to afford time for NMFS to consider the comments. NMFS submitted the final biological opinion to the NRC on January 30, 2013 (NMFS 2013), which concluded the formal consultation in accordance with 50 CFR 402.14(l).

After reviewing the proposed action, the status of the species, the environmental baseline, the effects of the action, and the cumulative effects including climate change, the biological opinion (NMFS 2013) concludes that

> [T]he continued operation of Indian Point Unit 2 is likely to adversely affect but is not likely to jeopardize the continued existence of shortnose sturgeon or the New York Bight, Gulf of Maine or Chesapeake Bay DPS [distinct population segments] of Atlantic sturgeon. It is also NMFS' biological opinion that the continued operation of Indian Point Unit 3 is likely to adversely affect but is not likely to jeopardize the continued existence of shortnose sturgeon or the New York Bight, Gulf of Maine or Chesapeake Bay DPS of Atlantic sturgeon. No critical habitat is designated in the action area; therefore, none will be affected by the proposed actions.

The biological opinion (NMFS 2013, page 126) finds that the "Hudson River population of shortnose sturgeon has experienced an increasing trend and is stable at high levels" and that renewal of the operating licenses would "not change the status or trend of the Hudson River population of shortnose sturgeon or the species as a whole" (NMFS 2013, page 119). It also finds that license renewal would "not change the status or trend of the Hudson River population of Atlantic sturgeon or the status and trend of the NYB DPS [New York Bight Distinct Population Segment] as a whole" (NMFS 2013, page 125).

The 2013 biological opinion includes an ITS that applies to both shortnose and Atlantic sturgeon impinged at IP2 and IP3 for both the remainder of the present license terms and the possible renewed license terms (NMFS 2013, page 127). The ITS (NMPF 2013 pp 130) exempts the following take (injure, kill, capture or collect) as described below:

- A total of two dead or alive shortnose sturgeon (injure, kill, capture or collect) and 2 dead or alive New York Bight DPS Atlantic sturgeon (injure, kill, capture or collect) impinged at the Unit 1 intake screens from now until the IP2 proposed renewed operating license would expire on September 28, 2033.

- A total of 395 dead or alive shortnose sturgeon (injure, kill, capture or collect) and 269 New York Bight DPS Atlantic sturgeon (injure, kill, capture or collect) impinged at Unit 2 intakes (Ristroph screens) from now until the IP2 proposed renewed operating license would expire on September 28, 2033.

- A total of 167 dead or alive shortnose sturgeon (injure, kill, capture or collect) and 145 dead or alive New York Bight DPS Atlantic sturgeon (injure, kill, capture or collect) impinged at the Unit 3 intakes (Ristroph screens) from now until the IP3 proposed renewed operating license would expire on December 12, 2035.

- All shortnose sturgeon with body widths greater than 3" impinged at the IP1, IP2 and IP3 trash racks (capture or collect).

- All Atlantic sturgeon with body widths greater than 3" impinged at the IP1, IP2 and IP3 trash racks (capture or collect). These Atlantic sturgeon will originate from the New York Bight (92), Gulf of Maine (6% ~~percent~~) and Chesapeake Bay DPSs (2% ~~percent~~).

NMFS (2013, pages 130-131) would consider the ITS to be exceeded if any one of 16 conditions occurs, each of which specifies the species and population of impinged fish, the number impinged, the generating unit involved, the location of impingement (intake screens or trash racks), and a time period. The ITS states (NMFS 2013, pages 132-133) that Entergy must comply with the following reasonable and prudent measures that NMFS finds necessary or appropriate to minimize and monitor impacts of incidental take of endangered shortnose and Atlantic sturgeon:

(1) A program to monitor the incidental take of shortnose and Atlantic sturgeon at the IP1, IP2 and IP3 intakes must be developed, approved by NMFS, and implemented as described in the Terms and Conditions [of the Biological Opinion]. This program must be implemented throughout the remaining duration of the existing IP2 and IP3 operating licenses as well as during the time IP2 and/or IP3 operate pursuant to the proposed renewed operating license(s).

(2) All live, incidentally taken shortnose and Atlantic sturgeon must be released back into the Hudson River at an appropriate location away from the intakes and thermal plume that does not pose additional risk of take, including death, injury, harassment, collection/capture.

(3) Any dead, incidentally taken shortnose or Atlantic sturgeon must be transferred to NMFS or an appropriately permitted research facility NMFS will identify so that a necropsy can be undertaken to attempt to determine the cause of death.

(4) A genetic sample must be taken of all incidentally taken Atlantic and shortnose sturgeon.

(5) All incidental takes of shortnose and Atlantic sturgeon associated with the Indian Point facilities and any shortnose or Atlantic sturgeon sightings in the action area must be reported to NMFS.

The ITS also contains eight specific, non-discretionary "terms and conditions" that implement the reasonable and prudent measures and outline required reporting and monitoring requirements. Entergy must comply with, and the NRC must ensure through enforceable terms of the existing and renewed licenses of IP2 and IP3 that Entergy does comply with, the terms and conditions of the ITS (NMFS 2013, pages 133-138). NMFS further identifies (NMFS 2013, pages 138-140) seven discretionary conservation recommendations that it recommends the NRC consider, and identifies the conditions for reinitiation of consultations.

4.5 Conclusion for Aquatic Special Status Species

In addition to the WOE information provided in Table 4–4, the staff examined the new information from the ESA Section 7 consultations with NMFS to determine the level of impact resulting from license renewal of IP2 and IP3 for the purposes of NEPA. Because NMFS (2013) finds that license renewal would not change the status or trend of the Hudson River population of shortnose sturgeon or the species as a whole, the NRC finds that the level of impact would be SMALL for this species. For Atlantic sturgeon, NMFS finds that license renewal would not change the status or trend of the Hudson River population of Atlantic sturgeon or the status and trend of the NYB DPS as a whole. NMFS (2013) calculates that the highest observed annual impingement of Atlantic sturgeon at the traveling screens would represent about 0.5 percent of the Hudson River origin juveniles. This potential reduction would not be observable or noticeable through any population study. Therefore, the staff finds that the level of impact would be SMALL for Atlantic sturgeon. Furthermore, development and implementation of an appropriate monitoring plan for these species at IP2 and IP3 would help ensure protection of these species. In addition, license renewal for IP2 and IP3 would be subject to the terms and conditions of the ITS as stated by NMFS. After assessing this new information, the staff finds that the level of impact for aquatic special status species would be SMALL.

5.0 REFERENCES

References that appear with an Agencywide Documents Access and Management System (ADAMS) accession number can be accessed through the U.S. Nuclear Regulatory Commision's (NRC's) Web-based ADAMS at the following URL: http://adams.nrc.gov/wba/.

10 CFR Part 51. *Code of Federal Regulations*, Title 10, *Energy*, Part 51, "Environmental protection regulations for domestic licensing and related regulatory functions."

50 CFR 402. *Code of Federal Regulations*, Title 50, *Wildlife and Fisheries*, Part 402, "Interagency cooperation—Endangered Species Act of 1973, as amended."

77 FR 5880. National Oceanic and Atmospheric Administration. "Endangered and Threatened Wildlife and Plants; Final Listing Determinations for Two Distinct Population Segments of Atlantic Sturgeon (*Acipenser oxyrinchus oxyrinchus*) in the Northeast." *Federal Register* 77(24):5880-5912. February 6, 2012.

77 FR 5914. National Oceanic and Atmospheric Administration. "Endangered and Threatened Wildlife and Plants; Final Listing Determinations for Two Distinct Population Segments of Atlantic Sturgeon (*Acipenser oxyrinchus oxyrinchus*) in the Southeast." *Federal Register* 77(24):5914-5982. February 6, 2012.

Endangered Species Act of 1973. 16 U.S.C. 1531, et seq.

[AKRF] AKRF, Inc. 2011a. *Review of Estimates of Numbers Entrained Presented in NRC 2009 EFH Assessment and 2010 FSEIS.* ADAMS Accession No. ML11286A140.

[AKRF] AKRF, Inc. 2011b. *Technical Review of FSEIS for Indian Point Nuclear Generating Unit Nos. 2 and 3; Sections 4.1.1–4.1.3 and Appendices H and I.* Prepared for Entergy Nuclear Operations, Inc. Hanover, MD: AKRF, Inc. March 28, 2011. ADAMS Accession No. ML110980163.

AKRF, Inc.; Normandeau Associates, Inc.; ASA Analysis & Communications, Inc.; and LWB Environmental Services, Inc. 2012. *Atlantic Sturgeon and Shortnose Sturgeon Impingement at IPEC Units 2 and 3: Review of Historical Data, Projections of Impingement, and Assessment of the Condition of Impinged Sturgeon Upon Arrival at IPEC.* July 23, 2012. Prepared for Indian Point Energy Center, Buchanan, NY. ADAMS Accession No. ML12206A028.

Barnthouse, L., Mattson M., and Young J.. 2011. *Shortnose Sturgeon: A Technical Assessment Pursuant to the Endangered Species Act.* Prepared for Entergy Nuclear Operations, Inc.; Entergy Nuclear Indian Point 2, LLC; and Entergy Nuclear Indian Point 3, LLC. April 2011. ADAMS Accession No. ML11126A202.

[CHGEC et al.] Central Hudson Gas and Electric Corporation, Consolidated Edison Company of New York, Inc., New York Power Authority, and Southern Energy New York. 1999. *Draft Environmental Impact Statement for State Pollutant Discharge Elimination System Permits for Bowline Point, Indian Point 2 and 3, and Roseton Steam Electric Generating Stations.* December 1999. ADAMS Accession No. ML083400128.

Dadswell, MJ. 1979. Biology and population characteristics of the shortnose sturgeon, *Acipenser brevirostrum* LeSueur 1818 (Osteichthyes: Acipenseridae), in the Saint John River estuary, New Brunswick, Canada. Canadian Journal of Zoology 57:2186–2210.

Dadswell MJ, Taubert BD, Squiers TS, Marchette D, and Buckley J. 1984. Synopsis of biological data on shortnose sturgeon, *Acipenser brevirostrum* LeSueur 1818. NOAA Technical

References

Report NMFS-14. Washington, DC: National Marine Fisheries Service. October 1984. 45 pp. Available at <http://www.nmfs.noaa.gov/pr/pdfs/species/shortnosesturgeon_biological_ data.pdf> (accessed January 12, 2012).

[Entergy] Entergy Nuclear Operations, Inc. 2007. Letter from F. Dacimo, Vice President of Operations License Renewal, Entergy, to NRC Document Control Desk. Subject: Supplement to License Renewal Application—Environmental Report References. December 20, 2007. ADAMS Accession Nos. ML080080199; Long River Survey: ML11279A044; and Fall Shoals Survey: ML080080291, ML080080298, ML080080306.

[Entergy] Entergy Nuclear Operations, Inc. 2009. Letter from F. Dacimo, Vice President of License Renewal, to Document Control Desk, NRC. Subject: Transmission of additional requested information regarding sturgeon impingement data. July 1, 2009. ADAMS Accession No. ML091950345.

[Entergy] Entergy Nuclear Operations, Inc. 2011a. Letter from F. Dacimo, Vice President of License Renewal, Entergy, to A. Stuyvenberg, Project Manager, NRC. Subject: Endangered Species Act consultation for Indian Point Nuclear Generating Unit Nos. 2 and 3. March 1, 2011. ADAMS Accession No. ML110670270.

[Entergy] Entergy Nuclear Operations, Inc. 2011b. Letter from F. Dacimo, Vice President of Operations License Renewal, to B. Holian, Division Director, NRC. Subject: Comments on final supplemental environmental impact statement for Indian Point Nuclear Generating. March 29, 2011. ADAMS Accession No. ML110980073.

[Entergy] Entergy Nuclear Operations, Inc. 2011c. Letter from F. Dacimo, Vice President of License Renewal, Entergy, to A. Stuyvenberg, Project Manager, NRC. Subject: Endangered Species Act consultation for Indian Point Nuclear Generating Unit Nos. 2 and 3. April 28, 2011. ADAMS Accession No. ML11125A071.

[Entergy] Entergy Nuclear Operations, Inc. 2011d. Letter from F. Dacimo, Vice President of License Renewal, Entergy, to NRC Document Control Desk. Subject: License renewal thermal study documents for Indian Point Unit Nos. 2 and 3. June 29, 2011. ADAMS Accession No. ML11189A026.

[Entergy] Entergy Nuclear Northeast. 2011e. Letter from Fred Dacimo, Vice President, License Renewal, to Andrew Stuyvenberg, NRC Environmental Project Manager, NRC, and Patricia Kurkul, Regional Administrator, National Marine Fisheries Service - Northeast Region. Subject: Endangered Species Act Consultation, Indian Point Nuclear Generating Unit Nos. 2 & 3, Docket Nos. 50-247 and 50-286, License Nos. DPR-26 and DPR-64. April 28, 2011. ADAMS Accession No. ML11126A202.

[Entergy] Entergy Nuclear Northeast. 2012. Letter from Fred Dacimo, Vice President, License Renewal, to David Wrona, Branch Chief, Projects Branch 2, Division of License Renewal, NRC. Subject: Endangered Species Act Consultation, Indian Point Nuclear Generating Unit Nos. 2 & 3, Docket Nos. 50-247 and 50-286, License Nos. DPR-26 and DPR-64. NL-12-043. March 7, 2012. ADAMS Accession No. ML12074A116.

[Goodwin Proctor] Goodwin Proctor LLP. 2011. Letter from Elise N. Zoli, Goodwin Proctor LLC to David Wrona, Chief, Projects Branch 2, Division of License Renewal, U.S. Nuclear Regulatory Commission. Subject: Indian Point License Renewal-Entergy's Comments on NMFS' Essential Fish Habitat Consultation Correspondence. September 30, 2011. ADAMS Accession No. ML11286A140.

Heidt AR, and Gilbert RJ. 1978. The shortnose sturgeon in the Altamaha River drainage, Georgia. In: RR Odum and L Landers, editors. *Proceedings of the Rare and Endangered*

Wildlife Symposium. Georgia Department of Natural Resources, Game and Fish Division, Technical Bulletin WL-4. pp. 54–60.

Mendelsohn D, Swanson C, and Crowley D. 2011. *Part 1 of Response to the NYSDEC Staff Review of the 2010 Field Program and Modeling Analysis of the Cooling Water Discharge from the Indian Point Energy Center.* Prepared for Entergy Nuclear Indian Point 2, LLC and Entergy Nuclear Indian Point 3, LLC. South Kingstown, RI: Applied Science Associates, Inc. March 31, 2011. 13 pp. Available at <http://www.dec.ny.gov/docs/permits_ej_operations_pdf/indnptpart1resp.pdf> (accessed January 11, 2012).

[NMFS] National Marine Fisheries Service. 2009. Letter from M. Colligan, Northeast Assistant Regional Administrator for Protected Resources, NMFS, to D. Wrona, Branch Chief, NRC. Subject: Biological assessment for license renewal of Indian Point Nuclear Generating Unit Nos. 2 and 3. February 24, 2009. ADAMS Accession No. ML090820316.

[NMFS] National Marine Fisheries Service. 2011. Letter from P. Kurkul, Northeast Regional Administrator, to D. Wrona, Branch Chief, NRC. Subject: Reply to biological assessment for license renewal of the Indian Point Nuclear Generating Unit Nos. 2 and 3. February 16, 2011. ADAMS Accession No. ML110550751.

[NMFS] National Marine Fisheries Service. 2011a. Letter from P. Kurkul, Northeast Assistant Regional Administrator for Protected Resources, NMFS, to D. Wrona, Branch Chief, NRC. Subject: Extension of consultation period—license renewal of the Indian Point Nuclear Generating Plant, Unit Nos 2 and 3. March 16, 2011. ADAMS Accession No. ML110830578.

[NMFS] National Marine Fisheries Service. 2011b. Letter from P. Kurkul, Northeast Assistant Regional Administrator for Protected Resources, NMFS, to D. Wrona, Branch Chief, NRC. Subject: Draft biological opinion for license renewal of Indian Point Nuclear Generating Unit Nos. 2 and 3. August 26, 2011. ADAMS Accession No. ML11249A012.

[NMFS] National Marine Fisheries Service. 2011c. E-mail from J. Crocker, Fisheries Biologist, to A. Stuyvenberg, Project Manager, NRC. Subject: Schedule for biological opinion (revised proposal). September 19, 2011. ADAMS Accession No. ML11300A037.

[NMFS] National Marine Fisheries Service. 2011d. Letter from P. Kurkul, Northeast Assistant Regional Administrator for Protected Resources, NMFS, to D. Wrona, Branch Chief, NRC. Subject: Biological opinion for relicensing of Indian Point Nuclear Generating Unit Nos. 2 and 3. October 14, 2011. ADAMS Accession No. ML11290A232.

[NMFS] National Marine Fisheries Service. 2011e. *Biological Opinion for Relicensing—Indian Point Nuclear Generating Station F/NER/2009/00619.* October 14, 2011. ADAMS Accession No. ML11290A231.

[NMFS] NOAA's National Marine Fisheries Service. 2013. Endangered Species Act Section 7 Consultation Biological Opinion: Continued Operation of the Indian Point Nuclear Generating Station, Units 2 and 3, Pursuant to Existing and Proposed Renewed Operating Licenses, NER-2012-25252. January 30, 2013. ADAMS No. ML13032A256.

[NRC] U.S. Nuclear Regulatory Commission. 2008a. *Biological Assessment for License Renewal of Indian Point Nuclear Generating Unit Nos. 2 and 3.* Appendix E to draft NUREG-1437, Supplement 38. 16 pp. ADAMS Accession No. ML083540614.

[NRC] U.S. Nuclear Regulatory Commission. 2008b. Letter from D. Wrona, Branch Chief, NRC, to M. Colligan, Northeast Assistant Regional Administrator for Protected Resources, NMFS. Subject: Biological assessment for license renewal of Indian Point Nuclear Generating Unit Nos. 2 and 3. December 22, 2008. ADAMS Accession No. ML083450723.

References

[NRC] U.S. Nuclear Regulatory Commission. 2009. E-mail from D. Logan, Aquatic Ecologist, NRC, to J. Crocker, NMFS. Subject: Indian Point Section 7—new data from Entergy. August 10, 2009. ADAMS Accession No. ML092220524.

[NRC] U.S. Nuclear Regulatory Commission. 2010a. Generic Environmental Impact Statement for License Renewal of Nuclear Plants: Supplement 38, Regarding Indian Point Nuclear Generating Unit Nos. 2 and 3. Washington, DC: NRC. NUREG–1437, Supp. 38. December 2010. ADAMS Accession No. ML103270072.

[NRC] U.S. Nuclear Regulatory Commission. 2010b. Letter from D. Wrona, Branch Chief, NRC, to M. Colligan, Northeast Assistant Regional Administrator for Protected Resources, NMFS. Subject: Revised biological assessment for license renewal of the Indian Point Nuclear Generating Unit Nos. 2 and 3. December 10, 2010. ADAMS Accession No. ML102990043.

[NRC] U.S. Nuclear Regulatory Commission. 2011a. Summary of public meeting held on April 18, 2011, between NRC and Entergy to discuss Entergy's shortnose sturgeon and Atlantic sturgeon data at Indian Point Nuclear Generating Unit Nos. 2 and 3. April 22, 2011. ADAMS Accession No. ML111090905.

[NRC] U.S. Nuclear Regulatory Commission. 2011b. E-mail from A. Stuyvenberg, Project Manager, NRC, to J. Crocker, Fisheries Biologist, NMFS. Subject: Indian Point thermal information available on New York DEC website. June 16, 2011. ADAMS Accession No. ML11167A108.

[NRC] U.S. Nuclear Regulatory Commission. 2011c. Letter from L. Bauer, Acting Branch Chief, NRC, to M. Colligan, Northeast Assistant Regional Manager for Project Resources, NMFS. Subject: Supplement to revised biological assessment for license renewal of Indian Point Nuclear Generating Unit Nos. 2 and 3. July 26, 2011. ADAMS Accession No. ML11203A100.

[NRC] U.S. Nuclear Regulatory Commission. 2011d. Summary of telephone conference calls held on Jun. 20, 22, and 29, 2011, regarding the ongoing Endangered Species Act consultation for the proposed Indian Point Nuclear Generating Unit Nos. 2 and 3 license renewal. July 29, 2011. ADAMS Accession No. ML11201A306.

[NRC] U.S. Nuclear Regulatory Commission. 2011e. E-mail from A. Stuyvenberg, Project Manager, NRC, to J. Crocker, Fisheries Biologist, NMFS. Subject: NRC staff comments on draft BO for proposed Indian Point license renewal. September 6, 2011. ADAMS Accession No. ML11249A210.

[NRC] U.S. Nuclear Regulatory Commission. 2011f. E-mail from A. Stuyvenberg, Project Manager, NRC, to J. Crocker, Fisheries Biologist, NMFS. Subject: FW: Entergy comments on draft BO. September 6, 2011. ADAMS Accession No. ML11249A145.

[NRC] U.S. Nuclear Regulatory Commission. 2011g. Letter from D. Wrona, Branch Chief, NRC, to P. Kurkul, Northeast Assistant Regional Administrator for Protected Resources, NMFS. Subject: NMFS Letter dated August 26, 2011, regarding the Endangered Species Act, Section 7 consultation for the proposed license renewal of Indian Point Nuclear Generating Unit Nos. 2 and 3. September 20, 2011. ADAMS Accession No. ML11259A018.

[NRC] U.S. Nuclear Regulatory Commission. 2011h. Summary of telephone conference calls held on March 9 and March 11, 2011, regarding the ongoing Endangered Species Act consultation for the proposed Indian Point Nuclear Generating Unit Nos. 2 and 3 license renewal. April 14, 2011. ADAMS Accession No. ML11089A031.

[NRC] U.S. Nuclear Regulatory Commission. 2012a. Letter from J.J. Susco, Acting Chief, Environmental Review and Guidance Update Branch, Division of License Renewal to Patricia A. Kurkul, Northeast Regional Administrator, National Marine Fisheries Service. Subject: Request to Reinitiate Section 7 Consultation for the Indian Point Nuclear Generating Unit Nos. 2 and 3 Due to Listing of Atlantic Sturgeon. May 16, 2012. ADAMS Accession No. ML12100A082.

[NRC] U.S. Nuclear Regulatory Commission. 2012b. E-mail from Dennis Logan, Aquatic Biologist, NRC, to Julie Crocker, Fisheries Biologist, NMFS. Subject: NRC and Entergy comments on NMFS's draft Indian Point biological opinion. November 9, 2012. ADAMS No. ML12314A415.

[NYSDEC] New York Department of Environmental Conservation. 2010. Letter from W.R. Adriance, Chief Permit Administrator, NYSDEC to D. Gray, Entergy. Subject: Joint Application for CWA §401 Water Quality Certification, NRC License Renewal—Entergy Nuclear Indian Point Units 2 and 3, DEC Nos.: 3-5522-00011/00030 (IP2) and 3-5522-00105/00031 (IP3), Notice of Denial. April 2, 2010. Available at <http://www.dec.ny.gov/docs/permits_ej_operations_pdf/ipdenial4210.pdf> (accessed January 11, 2012).

[NYSDEC] New York Department of Environmental Conservation. 2011. Letter from M.D. Sanza, Assistant Counsel, NYSDEC to Administrative law Judges M. Villa and D.P. O'Connell, NYSDEC Office of Hearings and Mediation Services. Subject: Entergy Indian Point Nuclear Units 2 and 3 SPDES Permit Renewal/401 WQC Application Proceedings; DEC Staff's Review of Thermal Information. May 16, 2011. Available at <http://www.dec.ny.gov/docs/permits_ej_operations_pdf/indnptsanzaltr.pdf> (accessed January 4, 2012).

Swanson C, Kim Y, Mendelsohn D, Crowley D, and Mattson M. 2010. *Preliminary Analysis of Hudson River Thermal Data.* South Kingstown, RI: Applied Sciences, Inc. and Bedford, NH: Normandeau Associates, Inc. February 10, 2010. 20 pp. Available at < http://www.dec.ny.gov/docs/permits_ej_operations_pdf/elecbdrexhj-k.pdf> (accessed January 11, 2012).

Swanson C, Mendelsohn D, Cohn N, Crowley D, Kim Y, Decker L, and Miller L. 2011a. *2010 Field Program and Modeling Analysis of the Cooling Water Discharge From the Indian Point Energy Center.* Prepared for Indian Point Entergy Center, Buchanan, NY. South Kingstown, RI: Applied Science Associates, Inc. January 31, 2011. 132 pp. Available at: <http://www.dec.ny.gov/docs/permits_ej_operations_pdf/indnpthrmlrpt.pdf> (accessed January 11, 2012). ADAMS Accession No. ML11189A026.

Swanson C, Crowley D, Kim Y, Cohn N, and Mendelsohn D. 2011b. *Part 2 of Response to the NYSDEC Staff Review of the 2010 Field Program and Modeling Analysis of the Cooling Water Discharge from the Indian Point Energy Center.* Prepared for Entergy Nuclear Indian Point 2, LLC and Entergy Nuclear Indian Point 3, LLC. South Kingstown, RI: Applied Science Associates, Inc. March 31, 2011. 27 pp. Available at <http://www.dec.ny.gov/docs/permits_ej_operations_pdf/indnptpart2resp.pdf> (accessed January 11, 2012).

Ziegeweid JR, Jennings CA, and Peterson DL. 2008a. Thermal maxima for juvenile shortnose sturgeon acclimated to different temperatures. Environmental Biology of Fish 3:299–307.

Ziegeweid JR, Jennings CA, Peterson DL, and Black MC. 2008b. Effects of salinity, temperature, and weight on the survival of young-of-year shortnose sturgeon. Transactions of the American Fisheries Society 137:1490–1499.

6.0 LIST OF PREPARERS

Members of the NRC's Office of Nuclear Reactor Regulation prepared this SEIS with assistance from other NRC organizations, as well as contract support from the Pacific Northwest National Laboratory. Table 6–1 identifies each contributor's name, affiliation, and function or expertise.

Table 6–1. List of Preparers

Name	Affiliation	Function or Expertise
NRC		
Jeremy Susco	Nuclear Reactor Regulation	Branch Chief
David Wrona	Nuclear Reactor Regulation	Branch Chief
Melanie Wong	Nuclear Reactor Regulation	Branch Chief
Michael Wentzel	Nuclear Reactor Regulation	Project Manager
Lois James	Nuclear Reactor Regulation	Project Manager
Kimberly Green	Nuclear Reactor Regulation	Project Manager
Dennis Logan	Nuclear Reactor Regulation	Ecology
Briana Balsam	Nuclear Reactor Regulation	Ecology
Contractor		
Valerie Cullinan	Pacific Northwest National Laboratory	Statistics, Ecology
Jeffrey Ward	Pacific Northwest National Laboratory	Ecology

APPENDIX A

COMMENTS RECEIVED ON THE DRAFT SUPPLEMENT TO THE FSEIS
FOR LICENSE RENEWAL OF INDIAN POINT UNITS 2 AND 3

COMMENTS RECEIVED ON THE DRAFT SUPPLEMENT

On June 26, 2012, the U.S. Nuclear Regulatory Commission (NRC) staff issued the draft supplement to "Generic Environmental Impact Statement for License Renewal of Nuclear Plants Regarding Indian Point Nuclear Generating Unit Nos. 2 and 3, Final Report" (NUREG-1437, Supplement 38, Volume 4, referred to as the draft supplement to the FSEIS) to Federal, tribal, state, and local government agencies and interested members of the public for comment in accordance with 10 CFR 51.92(f)(1). The U.S. Environmental Protection Agency (EPA) issued its Notice of Availability on July 6, 2012 (77 FR 40036). The public comment period ended on August 20, 2012. As part of the process to solicit public comments on the draft supplement to the FSEIS, the NRC staff did the following:

- placed a copy of the draft supplement to the FSEIS at the Field Library in Peekskill, New York, the White Plains Public Library in White Plains, New York, and the Henrick Hudson Free Library in Montrose, New York;

- made the draft supplement to the FSEIS available in the NRC's Public Document Room in Rockville, Maryland;

- placed a copy of the draft supplement to the FSEIS on the NRC website at http://www.nrc.gov/reading-rm/doc-collections/nuregs/staff/sr1437/supplement38/v4/;

- provided a copy of the draft supplement to the FSEIS to any member of the public that requested one;

- sent copies of the draft supplement to the FSEIS to certain Federal, tribal, state, and local government agencies;

- filed the draft supplement to the FSEIS with the EPA; and

- published a notice of availability of the draft supplement to the FSEIS in the *Federal Register* on July 6, 2012 (77 FR 40092).

During the public comment period, the NRC staff received comments from eight individuals or groups. Each comment letter is part of the docket file for the IP2 and IP3 license renewal application, all of which are accessible in the NRC's Agencywide Documents Access Management System (ADAMS). ADAMS is accessible at http://www.nrc.gov/reading-rm/adams.html. Table A–1 lists each individual that provided a comment during the comment period, and their assigned correspondence identification number. The NRC staff reviewed and assigned each comment within each comment letter a specific comment identification number consisting of the correspondence identification number and a number associated with the sequential order of the comment within the specific document. Table A–2 lists the comments, grouped by category, and where the comment and response can be found within this appendix.

Table A–1. Individuals Providing Comments During the Comment Period

Commenter	Affiliation (if stated)	Comment Source (ADAMS Accession #)	Correspondence ID
Brancato, Deborah	Riverkeeper, Inc.	Letter ML12236A207	001
Bullard, John	National Marine Fisheries Service (NMFS)	Letter ML12230A106	002
Dacimo, Fred	Entergy Nuclear Operations, Inc.	Letter ML12244A002	003
Kremer, Arthur	New York Affordable Reliable Electricty Alliance	E-Mail ML12234A093	004
McTiernan, Edward	New York State Department of Environmental Conservation	Letter ML12235A149	005
Mitchell, Judy-Ann	U.S. Environmental Protection Agency (EPA)	Letter ML12244A003	006
Raddant, Andrew	U.S. Department of the Interior (DOI)	Letter ML12235A410	007
Sipos, John	New York State Office of the Attorney General	Letter ML12235A409	008

Table A–2. Comments by Category

Comment Category	Page	Commenter (Comment ID)
Aquatic	A-4	• Brancato, Deborah (001-1) (001-2) (001-3)
	A-7	• Dacimo, Fred (003-1)
	A-21	• McTiernan, Edward (005-1)
	A-24	• Mitchell, Judy-Ann (006-1)
Endangered Species	A-8	• Brancato, Deborah (001-4)
General	A-16	• Bullard, John (002-1)
	A-19	• Kremer, Arthur (004-1)
	A-26	• Raddant, Andrew (007-1)
License Renewal Process	A-14	• Brancato, Deborah (001-5)
	A-32	• Sipos, John (008-4)
Postulated Accidents	A-23	• McTiernan, Edward (005-2)
	A-27	• Sipos, John (008-1) (008-2) (008-3)

A.1   Public Comments and NRC Staff Responses

RIVERKEEPER.
NY's clean water advocate

August 20, 2012

SUBMITTED ELECTRONICALLY

Chief, Rules, Announcements, and Directives Branch
Division of Administrative Services
Office of Administration
Mail Stop: TWB-05-B01M
U.S. Nuclear Regulatory Commission
Washington, DC 20555-0001

Re: *Docket ID NRC-2008-0672 – Riverkeeper, Inc.'s Comments on the U.S. Nuclear Regulatory Commission's Generic Environmental Impact Statement for License Renewal of Nuclear Plants, Supplement 38, Vol. 4, Regarding Indian Point Nuclear Generating Unit Nos. 2 and 3, Draft Report for Comment, Docket Nos. 50-247 and 50-286 (June 2012)*

Dear Rules, Announcements, and Directives Branch Chief:

Riverkeeper, Inc. ("Riverkeeper") hereby respectfully submits the following comments on the U.S. Nuclear Regulatory Commission Staff's ("NRC Staff") Generic Environmental Impact Statement for License Renewal of Nuclear Plants, Supplement 38, Volume 4, Regarding Indian Point Nuclear Generating Unit Nos. 2 and 3, Draft Report for Comment (hereinafter referred to as "Draft FSEIS Supplement"). Notice of availability of, and opportunity to comment on, the Draft FSEIS Supplement was published on June 26, 2012.[1]

The NRC Staff initially issued a final supplemental environmental impact statement relating to the proposed license renewal of Indian Point in December 2010.[2] Based upon purported newly available information, the NRC Staff issued the above-referenced draft supplement to this final

---

[1] *See* Letter from David J. Wrona (NRC) to U.S. Environmental Protection Agency Office of Federal Activities NEPA Compliance Division EIS Filing Section, Re: Notice of Availability of Draft Supplement to Final Plant Specific Supplement 38 to the Generic Environmental Impact Statement for License Renewal of Nuclear Plants, Regarding Indian Point Nuclear Generating Unit Nos. 2 and 3 (June 26, 2012), ADAMS Accession No. ML12159A495 (indicating a comment period extending to August 20, 2012).

[2] *See* Generic Environmental Impact Statement for License Renewal of Nuclear Plants: Regarding Indian Point Nuclear Generating Unit Nos. 2 and 3 - Final Report, Main Report and Comment Responses (NUREG-1437, Supplement 38, Volumes 1-3), *available at*, http://www.nrc.gov/reading-rm/doc-collections/nuregs/staff/sr1437/supplement38/ (last visited Aug. 20, 2012).

www.riverkeeper.org • 20 Secor Road • Ossining, New York 10562 • t 914.478.4501 • f 914.478.4527

001-1A In the FSEIS, the staff addressed the Pisces 2009 comments on the draft SEIS and, in addition, modified its entrainment and impingement analysis methods in the FSEIS in response to new information and comments submitted on the DSEIS. The staff's new strength-of-connection analysis in the FSEIS did not contain elements about which Pisces had expressed concerns in the original analysis in the draft SEIS. Appendix A of the FSEIS presents the staff's responses, and the body of the FSEIS shows where text was changed. The staff agrees with the commenter that the new information incorporated in the supplement to the FSEIS changed the conclusions for one fish species but did not change the overall conclusion of the FSEIS regarding the effects of entrainment and impingement. The staff made no further changes in response to this comment and considers no further changes to be warranted.

001-1A

report.[3] In particular, NRC Staff's Draft FSEIS Supplement includes "corrections to impingement and entrainment data presented in the FSEIS and revised conclusions regarding thermal impacts" in light of new "thermal plume studies"; NRC Staff's draft supplement also provides "an update of the status of the NRC's consultation under section 7 of the Endangered Species Act with the National Marine Fisheries Service [NMFS] regarding shortnose sturgeon . . and Atlantic sturgeon."[4]

NRC Staff's Revised Analysis of Impingement and Entrainment Impacts at Indian Point

NRC Staff's Draft FSEIS Supplement includes a revised assessment of impingement and entrainment impacts based upon new information obtained from Entergy about impingement and entrainment field data units of measure.[5] However, NRC Staff's new analysis does not meaningfully alter the ultimate conclusion that the operation of Indian Point has, and will continue to have, a profoundly negative impact upon the aquatic ecology of the Hudson River.

Riverkeeper's expert biologist consultants at Pisces Conservation Ltd. ("Pisces"), who reviewed and commented upon NRC Staff's initial assessment of impingement and entrainment impacts at Indian Point,[6] have now also reviewed NRC Staff's new Draft FSEIS Supplement. Pisces has prepared a response to NRC Staff's new supplement, which is provided in support of the instant comments as Attachment A.[7] Pisces recognizes that Entergy's presentation of the data with incorrect units caused confusion and errors in the calculation of the number of organisms impinged and entrained at Indian Point.[8] However, Pisces points out that for most species, "the error in units cancelled themselves out," resulting in no change in NRC Staff's conclusions about the level of impact from impingement and entrainment at Indian Point on such species.[9] Pisces indicates that the only species greatly affected by NRC Staff's consideration of Entergy's "corrected" data was spottail shiner.[10] Pisces explains that even with this change, eight critical fish species continue to have a high strength of connection to the effects of Indian Point, and that Indian Point continues to have a "MODERATE" or "LARGE" impact on several fish species exhibiting this high level of connection.[11] Overall, NRC Staff's revised assessment did not meaningfully change the outcome of NRC Staff's analysis, or NRC Staff's ultimate conclusions about impingement and entrainment impacts caused by Indian Point.

[3] *See* Draft FSEIS Supplement at iii, ix, 1-2.

[4] *See id.*

[5] *See id.* at ix, 3-16.

[6] *See* Comment of Phillip Musegaas, Victor M. Tafur, and Deborah Brancato on Behalf of Riverkeeper, Inc. on Generic Environmental Impact Statement for License Renewal of Nuclear Plants, Supplement 38, Regarding Indian Point, Units 2 & 3 (March 18, 2009), ADAMS Accession No. ML090860983, at 5-9 (hereinafter "Riverkeeper Comments on Indian Point Dec. 2008 DSEIS").

[7] Pisces Conservation, Ltd. "Some notes on the Generic Environmental Impact Statement for License Renewal of Nuclear Plants - Supplement 38" (August 20, 2012) ("Attachment A – Pisces Memo").

[8] Attachment A – Pisces Memo at 1-2.

[9] *Id.*

[10] *Id.* at 2.

[11] *Id.*

2

Importantly, Pisces' original review of NRC Staff's draft assessment of entrainment and impingement at Indian Point revealed various deficiencies and inadequacies in the analysis.[12] As a result of such deficiencies, Pisces previously explained that the actual impact of Indian Point of various fish species was likely underestimated by NRC Staff.[13] NRC Staff's December 2010 FSEIS did not address Pisces' concerns or adequately recognize the devastating level of impact associated with the operation of Indian Point.[14] Likewise, NRC Staff's Draft FSEIS Supplement contains no analysis that addresses Pisces' original concerns. Nothing in NRC Staff's revised assessment alters the criticism articulated by Pisces relating to the flawed methodology employed by NRC Staff to determine impingement and entrainment impacts caused by Indian Point. Thus, for the reasons articulated in Pisces' original report concerning NRC Staff's environmental impact statement for the relicensing of Indian Point, NRC Staff's assessment remains fundamentally flawed and continues to misjudge the severity of impingement and entrainment at the plant.[15]

Indeed, the continued operation of Indian Point as proposed by Entergy, i.e., with the ongoing use of a once-through-cooling water intake structure, will result in significant impacts on an already stressed ecosystem.[16] This is simply not reflected in NRC Staff's Draft FSEIS

[12] *See* Riverkeeper Comments on Indian Point Dec. 2008 DSEIS, *supra* Note 6, at 5-9, and Exhibit A (P. A. Henderson & R. M. H. Seaby (Pisces Conservation Ltd), Comments relating to the Indian Point NRC draft EIS on the Cooling System (March 2009), at 1-9).

[13] *Id.*

[14] *See Generic Environmental Impact Statement for License Renewal of Nuclear Plants: Regarding Indian Point Nuclear Generating Unit Nos. 2 and 3* - Final Report, Main Report and Comment Responses (NUREG-1437, Supplement 38, Volume 1, at § 4.1.

[15] *See* Riverkeeper Comments on Indian Point Dec. 2008 DSEIS, *supra* Note 6, at 5-9, and Exhibit A (P. A. Henderson & R. M. H. Seaby (Pisces Conservation Ltd), Comments relating to the Indian Point NRC draft EIS on the Cooling System (March 2009), at 1-9).

[16] *See* Riverkeeper Comments on Indian Point Dec. 2008 DSEIS, *supra* Note 6 at 5-9, Exhibit A. The once-through cooling water system employed at Indian Point has a profound impact upon fish in the Hudson River. *See generally* Entrainment, Impingement and Thermal Impacts at Indian Point Nuclear Power Station, Pisces Conservation Ltd., November 2007, *available at*, http://www.riverkeeper.org/wp-content/uploads/2010/03/1397-PH-Henderson-Attachment-3-Expert-Cont-EC-1.pdf, at 44; *see id* at 4 ("Notably, "[t]he species for which entrainment mortality has been quantified form only a very small proportion of the total species present in the estuary. As was noted in the FEIS (page 53): '*Finally, although impingement and entrainment mortality is measured, it is typically measured only for several of the 140 species of fishes found in the Hudson. Information about the impact on the full suite of aquatic organisms is limited.*' The impact on other species is un-quantified and may be significant.") (emphasis in original); NYSDEC Fact Sheet, NY SPDES Draft Permit Renewal with Modification, Indian Point Electric Generating Station (Buchanan, NY – November 2003), at 2, Attachment B, page 1, http://www.dec.ny.gov/docs/permits_ej_operations_pdf/IndianPointFS.pdf ("Each year Indian Point Units 2 and 3 . . . cause the mortality of more than a billion fish from entrainment of various life stages of fishes through the plant and impingement of fishes on intake screens. . . . Thus, current losses of various life stages of fishes are substantial."); NYSDEC Hudson River Power Plants FEIS (June 25, 2003), at 2-3, *available at* http://www.dec.ny.gov/docs/permits_ej_operations_pdf/FEISHRPP1.pdf. DEC has characterized the destructive impacts associated with the operation of once-through cooling water intake structures as "comparable to habitat degradation; the entire natural community is impacted. . . . [I]mpingement and entrainment and warming of the water impact the entire community of organisms that inhabit the water column." NYSDEC Hudson River Power Plants FEIS (June 25, 2003), Public Comment Summary at 53-54, http://www.dec.ny.gov/docs/permits_ej_operations_pdf/FEISHRPPS.pdf (hereinafter "NYSDEC Power Plants FEIS

001-1C The New York State Department of Environmental Conservation (NYSDEC), not the NRC, regulates construction and operation cooling water intake structures under the Clean Water Act and both Federal and New York State regulations. At this time, NYSDEC and Entergy have made no decisions regarding installation and implementation of cylindrical wedgewire screens. The FSEIS presents a general discussion of wedgewire or fine-mesh screens in Section 4.1.5, Potential Mitigation Options. At this time, it is speculative whether cylindrical wedgewire screens will be installed and utilized at the facility. If NYSDEC and/or Entergy reach a decision that wedgewire screens will be installed, the NRC staff would consider whether that would require the NRC to reinitiate a Section 7 consultation with NMFS. The staff made no further changes to the supplement in response to this comment.

001-2 The staff's assessment of population trends in the FSEIS and this supplement to the FSEIS uses empirical monitoring study data from 1974 through 2005 (the most recent data then available), and its assessment of thermal impacts in the supplement uses data from a study that Entergy submitted to the NYSDEC in 2011. The staff's analyses based on these empirical studies take into account any changes to the fish populations in the Hudson River that have occurred over the years. The impingement and entrainment data used in the strength of connection analyses date from 1975 through 1990 and are the most recent impingement and entrainment data available.

Supplement -- NRC Staff unfailingly refuses to recognize the reality of the situation, and ascribe a realistic and accurate level of impact of entrainment and impingement on the aquatic ecology of the Hudson River. Notably, NRC Staff is content to review Entergy's proposal to operate Indian Point for an additional 20 years in a vacuum – that is, without adequately assessing Entergy's proposal to install and implement cylindrical wedgewire screens to purportedly reduce entrainment and impingement impacts, even though doing so will result in additional negative impacts to the aquatic ecology of the Hudson River, such as impacts to the river bottom.

In addition, NRC Staff's Draft FSEIS Supplement fails to address the fact that NRC Staff continues to rely on old data.[17] That is, all NRC Staff has done in the Draft FSEIS supplement is correct certain calculation errors with respect to decades-old data that is not necessarily reflective of current conditions, and does not take into account negative changes to the status of fish populations in the Hudson River that have occurred over the years.[18] This runs afoul of Council

Comment Summary"). Nearly 40 years of such degradation resulting from the use of once-through cooling at Indian Point has resulted in serious long-term impacts. Evidence indicates an increasingly unstable ecosystem and long-term declines for several signature Hudson River fish species. A Riverkeeper report released in May 2008, revealed that many Hudson River fish are in serious long-term decline. *See* The Status of Fish Populations and the Ecology of the Hudson, Pisces Conservation Ltd., April 2008, *available at*, http://www.riverkeeper.org/wp-content/uploads/2009/06/Status-of-Fish-in-the-Hudson-Pisces.pdf (hereinafter "Pisces 2008 Status of Hudson River Fish Report") (analyzing 13 "key" species of the Hudson River, and finding that 10 such species are in decline); *see also* NYSDEC Power Plants FEIS Comment Summary at 57 ("Several species of fish in the Hudson River estuary, such as American shad, white perch, Atlantic tomcod and rainbow smelt, have shown trends of declining abundance."). As DEC has stated, such "[d]eclines in the abundances of several species and changes in species composition raises concerns and questions regarding the health of the River's fish community." NYSDEC Power Plants FEIS Comment Summary at 58. With, by far, the largest water intake on the Hudson estuary, slaughtering hundreds of millions, and possibly over a billion aquatic organisms every year, the once-through cooling water intake structure at Indian Point has undoubtedly contributed to such decline, destabilization, and loss of aquatic resources. *See, e.g.,* Pisces 2008 Status of Fish Report at 37-38 ("The impact of Indian Point is the largest of several impacts from once-through cooling on the Hudson. When all the power plants are considered, the impact is large. . . "Tens- to hundreds-of-millions of eggs, larvae, and juvenile fishes of several species are killed per year for once-through users. The cumulative impact of multiple facilities substantially reduces the young-of-year (YOY) population for the entire river'. . . in some years these effects have been very large . . . between 33 – 79% reductions in Young of Year population. . . . Even if the power companies are not the sole cause of degradation of the Hudson River fish community, the loss of such high proportions of the fish populations must be important." (quoting NYSDEC Water Quality 2006 Report)); *see also* NYSDEC Power Plants FEIS Comment Summary at 58 (expressly recognizing that "[t]he millions of fish that are killed by power plants each year represent a significant mortality and are yet another stress on the River's fish community" that "must be taken into account when assessing these population declines."); NYS Governor's Office, Press Release, *With American Shad Stocks at Historically Low Levels, Governor Paterson Announces New Initiatives to Rebuild and Protect Hudson River Fisheries* (May 28, 2008), *available at*, http://www.state.ny.us/governor/press/press_0528082.html (last visited March 24, 2010) (In the context of announcing that Hudson River fisheries are in trouble, recognizing that "[t]he number of fish entering water intake pipes each year at the two Indian Point nuclear power plants alone is significant – over 1.2 billion fish eggs and larvae, including bay anchovy, striped bass, and Atlantic tomcod – with the vast majority dying during the process. Another 1.18 million fish per year become trapped against intake screens and likely die."). Entergy's insistence on relying upon an obsolete cooling technology and refusal to implement a far-superior closed-cycle system, would lead to two additional decades of enormous entrainment, impingement, and heat impacts on an already precarious ecosystem. This will lead to ongoing habitat degradation, and only further exacerbate the current decline and destabilization of Hudson River fish populations.

17 *See* Riverkeeper Comments on Indian Point Dec. 2008 DSEIS, *supra* Note 6, at 9.

18 *See generally supra* Note 16.

001-2 (cont) The staff has no reason to believe that the conditional impingement and entrainment mortality rate estimates based on those data would be different today, and the commenter presents no information to suggest that such rates have changed. The staff made no further changes to the supplement in response to this comment.

001-3 The NRC staff recognizes that the effects of the thermal discharge from IP2 and IP3 are the subject of an ongoing hearing in which new information and views are being assessed. At this time, the triaxial plume studies cited in the comment provide the best technical information available to the staff to assess possible thermal effects to aquatic resources. Those studies support the staff's conclusion. The staff did not change its conclusion about the level of impact as a result of this comment, although it recognizes that information that may become available in the future could lead to a different understanding at a later time. In addition, 10 CFR 51.92(a)(2) and (c) address preparation of a supplement to a final environmental impact statement for proposed actions that have not been taken, under the following conditions:

- There are new and significant circumstances or information relevant to environmental concerns and bearing on the proposed action or its impacts, or

- The NRC staff determines, in its opinion, that preparation of a supplement will further the purposes of NEPA.

001-2

001-3

on Environmental Quality ("CEQ") regulations implementing the National Environmental Policy Act ("NEPA"), which require that analyses in environmental impact statements have scientific integrity.[19]

For the foregoing reasons, NRC Staff's revised assessment of impingement and entrainment impacts caused by Indian Point remains inadequate.

NRC Staff's Revised Analysis of Thermal Discharge Impacts at Indian Point

NRC Staff's *Draft FSEIS Supplement* assesses "additional information from Entergy regarding the thermal plume" at Indian Point, and based upon that assessment, NRC Staff makes an allegedly "more informed conclusion regarding thermal impacts" of the plant.[20] In particular, NRC Staff reviewed a triaxial plume study that Entergy submitted to the New York State Department of Environmental Conservation ("DEC") as part of its State Pollutant Discharge Elimination System ("SPDES") permit renewal proceeding and Clean Water Act § 401 water quality certification denial appeal proceeding, correspondence between DEC and Entergy relating to this thermal study, and a DEC proposed determination that a 75-acre thermal mixing zone will provide reasonable assurance that the operation of Indian Point will comply with applicable regulations.[21] Whereas in NRC Staff's initial (December 2010) FSEIS, NRC Staff concluded that thermal impacts at Indian Point ranged from SMALL to LARGE, based on NRC Staff's review of the aforementioned new information, the *Draft FSEIS Supplement* indicates that now "NRC staff concludes that the impacts from heat shock to aquatic resources of the lower Hudson River would be SMALL."[22]

However, NRC Staff's changed conclusion is unjustified because Entergy's thermal study and DEC's *proposed* determination regarding the efficacy of a mixing zone, are highly disputed, namely by Riverkeeper, and currently the subject of ongoing adjudication. Indeed, Pisces' review of the thermal study after it was completed, resulted in detailed comments that outlined numerous concerns related to thermal impacts on aquatic ecology at Indian Point, and problems with Entergy's thermal study. These comments are included as Appendix 1 to Attachment A hereto.[23] Pisces' comments reveal that despite Entergy's thermal study and DEC's proposed mixing zone, thermal discharges from Indian Point will continue to pose a threat to the aquatic ecology of the river.[24] Moreover, Riverkeeper has vehemently opposed DEC's *proposal* to allow Entergy to operate with a mixing zone, raising numerous well-founded concerns about the legality and environmental efficacy of doing so. A copy of Riverkeeper's comments on DEC's

---

19 *See* 40 C.F.R. § 1502.24; *see also id.* § 1502.22.
20 Draft FSEIS Supplement at 17.
21 *Id.*
22 *Id.* at 20.
23 Attachment A – Pisces Memo, at Appendix 1 – Pisces Conservation Ltd, "Comments on the proposed Indian Point thermal mixing zone" (July 15, 2011).
24 *See id.* at pages 16 of 22 to 20 of 22.

5

proposed mixing zone at Indian Point is included with these comments as Attachment B.[25] This issue is currently the subject of *ongoing* adjudication in State proceedings before DEC. Thus, NRC Staff cannot simply indicate that "NYSDEC concluded that the results of the thermal plume studies provide reasonable assurance that the IP2 and IP3 discharge is in compliance with NYSDEC's water quality standards and criteria for thermal discharges," and thereby conclude that impacts of heat shock at Indian Point are SMALL.[26] Riverkeeper has raised valid concerns (that have yet to be fully resolved), which call into question Entergy's thermal study and DEC's proposed conclusions with respect to thermal impacts, and, in turn, NRC Staff's revised conclusions in the Draft FSEIS supplement.

For the foregoing reasons, NRC Staff's revised assessment of thermal impacts caused by Indian Point remains inadequate.

NRC Staff's "Update" on Endangered Species Act § 7 Consultations

NRC Staff's Draft FSEIS Supplement lastly discusses endangered species impacts at Indian Point.[27] First, NRC Staff discusses endangered shortnose sturgeon. In particular, NRC Staff revises its assessment of Indian Point's thermal impact on endangered shortnose sturgeon.[28] NRC Staff's revised conclusion "that the heated discharge resulting from the proposed IP2 and IP3 license renewal would have SMALL impacts on the shortnose sturgeon," is largely based on NRC Staff's consideration of Entergy's thermal study discussed above.[29] Riverkeeper respectfully submits that, due to the reasons discussed above regarding the potential ongoing thermal impacts from Indian Point, NRC Staff's conclusions are not entirely well-founded.[30] Moreover, Pisces specifically notes in relation to NRC Staff's Draft FSEIS Supplement that the NRC Staff's finding that there is a "SMALL" level of impact on endangered shortnose sturgeon at Indian Point requires verification.[31]

001-3

001-4A

---

[25] Letter from Mark Lucas (Riverkeeper) to Christopher M. Hogan (DEC), Re: *Entergy Nuclear Indian Point 2, LLC & Entergy Nuclear Indian Point 3, LLC Proposed Modification of Special Condition 7.b of SPDES Permit, DEC No. 3-5522- 00011/00004, SPDES No. NY-000472* (July 15, 2011) (Attachment B).

[26] Draft FSEIS Supplement at 20.

[27] *Id.* at 23-26.

[28] *Id.* at 23-24.

[29] *Id.*

[30] *See* Attachment A – Pisces Memo, at Appendix 1 – Pisces Conservation Ltd, "Comments on the proposed Indian Point thermal mixing zone" (July 15, 2011); Attachment B – Letter from Mark Lucas (Riverkeeper) to Christopher M. Hogan (DEC), Re: *Entergy Nuclear Indian Point 2, LLC & Entergy Nuclear Indian Point 3, LLC Proposed Modification of Special Condition 7.b of SPDES Permit, DEC No. 3-5522- 00011/00004, SPDES No. NY-000472* (July 15, 2011).

[31] *See id.* Moreover, it remains unclear whether, generally, the impact of Indian Point on shortnose sturgeon is "small." *See* Riverkeeper Comments on Indian Point Dec. 2008 DSEIS, at Appendix A (Pisces indicating that there is no reason to believe that an increasing population of shortnose sturgeon would lead to decrease in impingement and that with relatively rare fish, even a small number of impingement can have a big effect, and calling into question the ability of the NRC Staff to draw accurate conclusions based on obsolete data).

6

001-4A The staff addressed this comment in Section 4.0 of this supplement to the FSEIS, which has been revised to reflect the completion of consultations with NMFS on endangered species (including both shortnose sturgeon and Atlantic sturgeon), NMFS's biological opinion, and its issuance of an Incidental Take Statement for Indian Point Units 2 and 3.

NRC Staff's Draft FSEIS Supplement further memorializes the outcome of NRC Staff's Endangered Species Act ("ESA") section 7 consultations with NMFS concerning the impact of Indian Point on endangered shortnose sturgeon. Based on NRC Staff's mere summary of the sequence and outcome of the consultation process, NRC Staff has failed to comply with relevant regulations and guidance, which require *meaningful consideration* of the opinions and conclusions drawn by NMFS.[32] Indeed, NRC Staff does not indicate how NMFS' final biological opinion regarding endangered shortnose sturgeon affects it's NEPA-based analysis and conclusions regarding impacts to endangered resources. Instead, NRC Staff's discussion of the section 7 consultation process in the Draft FSEIS Supplement appears to be a purely opportunistic discussion, provided only because NRC Staff was issuing a draft supplement to address other issues anyway. This is further exemplified by NRC's treatment (i.e., acceptance) of the incomplete section 7 consultation process with respect to the newly endangered Atlantic sturgeon, as discussed below, more is required by controlling law and guidance.

In relation to Atlantic sturgeon, in light of the designation of this species as endangered on February 6, 2012, i.e., after the issuance of NRC Staff's December 2010 FSEIS, NRC Staff reinitiated section 7 consultation with NMFS.[33] However, NRC Staff simply indicates in the Draft FSEIS Supplement that it expects to carry on consultation procedures and "consider the results of that consultation, as appropriate."[34] This fails to assure compliance with NEPA, which requires full consideration of the consultation process and the opinions, conclusions, and recommendations of NMFS, *as part of* the NEPA assessment process. NRC Staff must include or consider NMFS' assessment, and issue a supplemental EIS to fully consider the outcome of the new section 7 consultation process. This must be accomplished prior to the finalization of the NEPA process concerning the proposed license renewal of Indian Point, and prior to any ultimate decision by the NRC regarding whether to relicense Indian Point.

In particular, the ESA provides that

> [e]ach Federal agency shall, in consultation with . . . the Secretary [of the Interior or Commerce as appropriate], insure that any action authorized, funded, or carried out by such agency . . . is not likely to jeopardize the continued existence of any endangered species or

[001-4A]

[001-4B]

001-4B The staff addressed this comment in Section 4.0 of this supplement to the FSEIS, which has been revised to reflect the completion of consultations with NMFS on endangered species (including both shortnose sturgeon and Atlantic sturgeon), NMFS's biological opinion, and its issuance of an Incidental Take Statement for Indian Point Units 2 and 3.

---

[32] See Endangered Species Consultation Handbook, *Procedures for Conducting Consultation and Conference Activities Under Section 7 of the Endangered Species Act*, U.S. Fish & Wildlife Service, National Marine Fisheries Service (March 1998), at 4-11, *available at* http://www.nmfs.noaa.gov/pr/pdfs/laws/esa_section7_handbook.pdf (hereinafter "NMFS Consultation Handbook"); 50 C.F.R. § 402.06(b); Interagency Cooperation – Endangered Species Act of 1973, as Amended, Final Rule, 51 Fed. Reg. 19,926 (1986); 50 C.F.R. § 402.15; ESA § 7(d), 16 U.S.C. § 1536(d).

[33] See id. at 26.

[34] See id. As of the date of these comments, the consultation process between NRC Staff and NMFS remains ongoing. See Correspondence from Amy Hull (NRC) to Mr. Daniel S. Morris (NMFS), Re: Response to Request for Additional Review Time for Endangered Species Act Section 7 Consultation at Indian Point Nuclear Generating Unit Nos. 2 and 3 (Aug. 17, 2012), ADAMS Accession No. ML12221A033 (approving a 60-day extension of the consultation process whereby NMFS agreed to provide NRC a draft biological opinion on October 22, 2012 for a two-week review and indicating that consultation will be completed by November 28, 2012).

7

Comment response on preceding page(s).

---

threatened species or result in the destruction or adverse modification of habitat of such species which is determined . . . to be critical."[35]

During formal consultation, NMFS must review all relevant information, evaluate the current status of the relevant listed species, evaluate the effects of the proposed action and cumulative effects on the listed species, formulate an opinion regarding whether the proposed action is likely to jeopardize the continued existence of the listed species, formulate discretionary conservation recommendations that would reduce or eliminate the impacts of the proposed action on listed species,[36] and formulate a statement concerning any incidental take of the listed species,[36] and formulate an opinion regarding any reasonable and prudent alternatives to the proposed project and reasonable and prudent measures that could be taken.[37] Formal consultation concludes when NMFS issues a "biological opinion" ("BO").[38] Once NMFS issues its BO, "the Federal agency shall determine whether and in what manner to proceed with the action in light of its section 7 obligations and the Service's biological opinion."[39]

In addition, NRC's NEPA-implementing regulations designate the impacts of license renewal on threatened or endangered species as a "Category 2" issue, i.e. one that requires site specific review during individual relicensing proceedings.[40] NRC's regulations acknowledge that "consultation with appropriate agencies would be needed at the time of license renewal *to determine whether threatened or endangered species are present and whether they would be adversely affected*."[41]

Federal regulations implementing the ESA contemplate coordination of the consultation process with environmental reviews pursuant to NEPA.[42] NMFS guidance on the consultation process further explains how

001-4B

---

[35] ESA § 7, 16 U.S.C. § 1536(a)(2).

[36] A statement from NMFS concerning any incidental take must specify the amount or extent of the impact, any "reasonable and prudent measures that the Director considers necessary or appropriate to minimize such impacts," and any "terms and conditions (including but not limited to, reporting requirements) that must be complied with by the Federal agency or any applicant to implement [such] measures." 50 C.F.R. § 402.14(i).

[37] *See* 50 C.F.R. § 402.14(g)

[38] *See id.* § 402.14(i).

[39] *Id.* § 402.15.

[40] *See* 10 C.F.R. Part 51, Table B-1 of Appendix B to Subpart A; GEIS § 3.9 ("Because compliance with the Endangered Species Act cannot be assessed without site-specific consideration of potential effects on threatened and endangered species, it is not possible to determine generically the significance of potential impacts to threatened and endangered species. This is a Category 2 issue.").

[41] 10 C.F.R. Part 51, Table B-1 of Appendix B to Subpart A (emphasis added).

[42] *See* 50 C.F.R. § 402.06(a) ("Consultation, conference, and biological assessment procedures under section 7 may be consolidated with interagency cooperation procedures required by other statutes, such as the National Environmental Policy Act. . . . The Service will attempt to provide a coordinated review and analysis of all environmental requirements.").

8

[F]ormal consultation and the Services' preparation of a biological opinion often involve coordination with the preparation of documents mandated by other environmental statutes and regulations, including . . . NEPA. . . . The Services should assist the action agency or applicant in *integrating the formal consultation process into their overall environmental compliance.*[43]

Pertinently, ESA regulations and the NMFS Consultation Handbook indicate that "[a]t the time the Final EIS is issued, *section 7 consultation should be completed*" and that "[t]he Record of Decision should *address the results of section 7 consultation*."[44] Indeed, only *after* the issuance of a BO can the Federal agency "determine whether and in what manner to proceed with the action in light of its section 7 obligations and the Service's biological opinion."[45] This settled and proper approach is demonstrated by numerous instances where ESA § 7 consultation processes were concluded well prior to the completion of a concurrent NEPA review process, and where a BO prepared by NMFS (or FWS) was incorporated into the final EIS and formed part of the basis for the Federal agency's final decision-making.[46]

[001-4B]

---

[43] NMFS Consultation Handbook, *supra* Note at 32, at p.4-11 (emphasis added); *see id.* ("A major concern of action agencies is often the timing of the consultation process in relation to their other environmental reviews. For example, since the time required to conduct formal section 7 consultation may be longer than the time required to complete preparation of NEPA compliance documents, the action agency should be encouraged to initiate informal consultation prior to NEPA public scoping. Biological assessments may be completed prior to the release of the Draft Environmental Impact Statement (DEIS) and formal consultation, if required, should be initiated prior to or at the time of release of the DEIS. Early inclusion of section 7 in the NEPA process would allow action agencies to share project information earlier and would improve interagency coordination and efficiency.").

[44] *Id.* (emphasis added); *see* 50 C.F.R. § 402.06(b) ("Where the consultation . . . has been consolidated with the interagency cooperation procedures required by other statutes such as NEPA . . . , the results should be included in the documents required by those statutes."); Interagency Cooperation – Endangered Species Act of 1973, as Amended, Final Rule, 51 Fed. Reg. 19926 (1986) (NMFS and the U.S. Fish and Wildlife Service ("FWS") jointly enacting regulations implementing the ESA, explaining that "the biological opinion *should be stated in the final environmental impact statement*") (emphasis added); *id.* (explaining that "[a] statement of the opinion may be a summary of its findings and conclusions" although "[t]he Service does feel that the entire opinion should be attached as an exhibit to the NEPA document if completion time permits.").

[45] 50 C.F.R. § 402.15; *see also* ESA § 7(d), 16 U.S.C. § 1536(d) (prohibiting agency action that forecloses formulation of reasonable measures/alternatives while consultation is ongoing).

[46] *See, e.g., National Parks & Conservation Ass'n v. U.S. Dep't of Transportation,* 222 F.3d 677, 679, 682 (9th Cir. 2000) (BA and BO prepared pursuant to ESA both incorporated into Federal agency's Final EIS, forming part of the basis for agency's informed decision, which satisfied NEPA); *Miccosukee Tribe of Indians of Fla. V. U.S. Army Corp. of Eng'rs,* 509 F. Supp. 2d 1288, 1294 (S.D. Fla. 2007) (Army Corp appending BO to final supplemental EIS and pointing to "years of consultation and cooperation with the FWS which preceded the FSEIS" to justify its environmental analysis; Court finding that "the analysis in the FSEIS, *including the attached BiOpp*, [biological opinion] is sufficient") (emphasis added); *Nw. Envtl. Advocates v. NMFS,* 2005 U.S. Dist. LEXIS 41828, *6 (W.D. Wash. 2005) (Federal agency "solicited comments on its draft FSEIS, *including the NMFS Biological Opinion.* After considering and responding to the public comments, the Corps issued its FSEIS"); *Seattle Audubon Society v. Lyons,* 871 F. Supp. 1291, 1305, 1314, 1320 (W. D. Wash. 1994) (FWS issued a biological opinion that was appended to the final EIS concerning a federal forest management plan, which formed part of basis for the Federal agency's final determinations).

9

001-4B

Since Atlantic sturgeon was listed after NRC Staff's issuance of the Indian Point license renewal FSEIS, there was no consultation process to be incorporated into the December 2010 FSEIS. However, this does not relieve NRC Staff of the obligation to ensure proper consideration of the now ongoing section 7 consultation procedures. NRC Staff's vague reference to potentially considering the outcome of the section 7 consultation process related to Atlantic sturgeon does not ensure that the impacts to this critical species will be adequately considered by NRC Staff in the Indian Point relicensing NEPA process. Indeed, there is no indication that NRC Staff's NEPA review will fully address the findings, conclusions, or recommendations of NMFS relating to endangered Atlantic sturgeon present in the Hudson River. Based on the pithy "update" provided in the NRC Staff's Draft FSEIS Supplement, it appears that NRC Staff may continue to rely on its own analysis, and not on the input to be provided by NMFS. While the Draft FSEIS Supplement recognizes that the consultation process remains open, NRC Staff did not address in any way how the very relevant, as yet unwritten BO by NMFS would factor into the NRC Staff's FSEIS or NRC Staff's final decision-making regarding the license renewal of Indian Point.

This renders NRC Staff's Draft FSEIS Supplement and NEPA process fundamentally flawed. NRC Staff's apparent position that completing the NEPA review related to the proposed relicensing of Indian Point prior to the completion of the ESA § 7 consultation process with NMFS concerning Atlantic sturgeon, runs contrary to the ESA, applicable regulations and guidance, and settled practice, as discussed above. NRC simply cannot arrive at final NEPA conclusions regarding impacts to endangered Atlantic sturgeon and, ultimately whether to recommend license renewal of Indian Point, without satisfying its ESA § 7 obligations and fully considering NMFS' prospective biological opinion.[47] Indeed, such a regulatory scheme is the only way to ensure adequate and appropriate consideration of impacts to endangered or threatened species, and thereby comply with basic tenets of NEPA. The fundamental purpose of NEPA is to "ensure[] that the agency, in reaching its decision, will have available, and will carefully consider, detailed information concerning significant environmental impacts" and to "guarantee[] that the relevant information will be made available to the larger audience that may also play a role in both the decisionmaking process and the implementation of that decision."[48] Thus, an EIS prepared pursuant to NEPA must be searching and rigorous, providing a "hard look" at the environmental consequences of the agency's proposed action.[49] It is impossible to conclude that NRC Staff's final determinations in the ultimate final FSEIS supplement can be

[47] *See* 50 C.F.R. § 402.15 (only *after* the issuance of a BO can the Federal agency "determine whether and in what manner to proceed with the action in light of its section 7 obligations and the Service's biological opinion."); *see also* ESA § 7(d), 16 U.S.C. § 1536(d) (prohibiting agency action that forecloses formulation of reasonable measures/alternatives while consultation is ongoing).

[48] *Entergy Nuclear Generation Co. and Entergy Nuclear Operations, Inc.* (Pilgrim Nuclear Power Station), LBP-06-23, 64 NRC 257, 277 (2006), quoting *Robertson v. Methow Valley Citizens Council*, 490 U.S. 332, 349 (1989); *see also Vermont Yankee Nuclear Power Corp. V. Natural Resources Defense Council*, 435 U.S. 519, 558 (1978) (explaining how NEPA seeks to ensure "a fully informed and well-considered decision"); *Nw. Envtl. Advocates v. NMFS*, 2005 U.S. Dist. LEXIS 41828, *6 (W.D. Wash. 2005) ("The processes established under NEPA focus the attention of both the government and the public on a proposed agency action, so that the environmental consequences can be studied prior to implementation of the proposed action, and so potential negative impacts can be avoided") (citing 40 C.F.R. § 1500.1(b); 40 C.F.R. § 1500.2(c); *Marsh v. Oregon Natural Resources Council*, 490 U.S. 360, 371 (1989); *Churchill County v. Norton*, 276 F.3d 1060, 1072-73 (9th Cir. 2001)).

[49] *Marsh*, 490 U.S. at 374,

10

001-4B

considered "fully-informed" and based on the requisite "hard look," if they are not informed by *any* feedback from the ESA § 7 consultation process related to Atlantic sturgeon (or if an additional supplement to the FSEIS is not prepared upon completion of the section 7 consultation process). Indeed, finalizing the NEPA process without the benefit of NMFS' assessment effectively ensures that NRC Staff's determinations regarding impacts to endangered species and the license renewal of Indian Point will not take into account any conclusions, findings, or recommendations of the consulting agency. This completely flouts the purpose of ESA § 7, which requires consultation with NMFS so as to inform the Federal agency's decision on the action to make certain that such action will not jeopardize any endangered species.[50]

For example, NMFS is charged with making an independent determination regarding whether the proposed action is likely to jeopardize any endangered species, making discretionary conservation recommendations to reduce or eliminate any impacts, determining whether a take permit is necessary, and formulating an opinion regarding any reasonable and prudent alternatives to the proposed project.[51] The opinions and recommendations from NMFS are highly critical given NRC Staff's continued reliance on outdated and/or incomplete information regarding impacts to Atlantic sturgeon.[52] NMFS' assessment will contain opinions that will necessarily inform the relevant concerns, including opinions and conclusions that may well differ from those of NRC Staff, and that logically should be considered before NRC Staff arrives at any final conclusions about impacts to endangered species and, in turn, whether license renewal of Indian Point is appropriate. Without the benefit of NMFS' BO (which will contain NMFS' position on the impacts of the activity, potential alternatives, mitigation measures, the necessity of obtaining a take permit, etc), NRC Staff does not have all of the information necessary to make the relevant findings regarding the license renewal of Indian Point. Failure to fully consider the section 7 consultation process related to Atlantic sturgeon will result in determinations by NRC Staff that do not adequately take into account adverse impacts on endangered species, which NMFS may find to be significant and "likely to jeopardize the continued existence" of such species.[53]

In sum, NRC Staff cannot draw final conclusions regarding the impact of Indian Point on Atlantic sturgeon in the Hudson River, or finalize the NEPA review process concerning the proposed license renewal of Indian Point, without a full and adequate consideration of the section 7 consultation process and input from NMFS. Notably, Pisces agrees that "[w]ithout more information an assessment for the Atlantic sturgeon is not possible."[54] A site specific assessment of environmental impacts of license renewal on Atlantic sturgeon is necessary for

---

[50] *See 16 U.S.C. § 1536(a)(2); 50 C.F.R. § 402.14(g).*

[51] *See 50 C.F.R. § 402.14(g).*

[52] *See, e.g.,* U.S. NRC, Biological Assessment for Reinitiation of Section 7 Consultation for the Indian Point Nuclear Generating Plant, Unit Nos. 2 and 3 Due to Listing of Atlantic Sturgeon, May 2012, ADAMS Accession No., ML12138A388, at 4, 10, Appendix A; *see also* Revised Biological Assessment of the Potential Effects on Federally Listed Endangered or Threatened Species from the Proposed Renewal of Indian Point Nuclear Generating Plant, Unit Nos. 2 and 3 (December 2010), ADAMS Accession No. ML102990046 (basing conclusions on "two-decade old impingement data and incomplete impingement mortality data.").

[53] *See 50 C.F.R. § 402.14(g)(4).*

[54] Attachment A – Pisces Memo at p.3 of 22.

11

NRC Staff to make informed conclusions in the FSEIS, and, in turn, informed recommendations regarding the appropriateness of relicensing Indian Point. Without meaningful consideration of NMFS' analysis pursuant to consultation procedures set forth by ESA § 7, the current findings in the FSEIS and Draft FSEIS Supplement in relation to impacts to endangered and threatened species lack proper foundation and are flawed and patently deficient.

For the foregoing reasons, NRC Staff's revised assessment of endangered species impacts caused by Indian Point remains inadequate.

NRC Cannot Issue Renewed Operating Licenses to Indian Point Unless and Until Entergy Obtains All Required and Necessary State Approvals and Certifications

Lastly, to the extent clarity is required notwithstanding the fact that the record is abundantly clear in the Indian Point license renewal proceeding, Riverkeeper reiterates the position that Entergy must obtain a new water quality certification pursuant to CWA § 401 prior to any license renewal for the plant. NRC Staff's December 2010 FSEIS acknowledged the ongoing nature of Entergy's appeal proceeding relating to NYSDEC's denial of Entergy's request for a CWA § 401 water quality certification.[55] In light of a recent United States Court of Appeals decision that was issued after the publication of NRC Staff's FSEIS, it may be useful to include in NRC Staff's supplemental NEPA document an explanation regarding the unequivocal obligation of the NRC to comply with CWA § 401, and the distinguishing nature of the recent court ruling; Riverkeeper's position is fully explained in a letter that was provided to the NRC on July 26, 2012, which is included with these comments as Attachment C.[56]

Notably, as NRC Staff has previously acknowledged in its initial FSEIS, Indian Point must receive a federal consistency determination from the State pursuant to the Coastal Zone Management Act[57] before NRC may issue operating licenses authorizing the operation of Indian Point Units 2 & 3 beyond their initial 40-year terms.[58] NRC may not issue a license renewal prior to the issuance of the federal consistency concurrence by the Department of State pursuant to 16 U.S.C. § 1456(3)(A), which requires that "[n]o license or permit shall be granted by the

001-4B

---

[55] Generic Environmental Impact Statement for License Renewal of Nuclear Plants, Supplement 38, Volume 1, Regarding Indian Point Nuclear Generating Unit Nos. 2 and 3, Docket Nos. 50-247 and 50-286 (December 2010), *available at,* http://pbadupws.nrc.gov/docs/ML1033/ML10350405.pdf, at xv ("Two state level issues (consistency with State water quality standards, and consistency with State coastal zone management plans) need not be resolved. On April2, 2010, the New York State Department of Environmental Conservation (NYSDEC) issued a Notice of Denial regarding the Clean Water Act Section 401 Water Quality Certification. Entergy has since requested a hearing on the issue, and the matter will be decided through NYSDEC's hearing process."); *see id.* at xvii-xviii, 1-8, 2-27, 4-8 to 4-9, 4-30, 8-3, 9-5, A-151.

[56] Letter from Deborah Brancato (Riverkeeper) to NRC Commissioners, Re: Entergy Nuclear Operations, Inc. (Indian Point Nuclear Generating Units 2 and 3), Docket Nos. 50-247-LR 50-286- LR (July 26, 2012) (Attachment C).

[57] 16 U.S.C. §§ 1451-1464.

[58] *See* Generic Environmental Impact Statement for License Renewal of Nuclear Plants: Regarding Indian Point Nuclear Generating Unit Nos. 2 and 3 - Final Report, Main Report and Comment Responses (NUREG-1437, Supplement 38, Volume 1), *available at,* http://pbadupws.nrc.gov/docs/ML1033/ML10350405.pdf, (last visited Aug. 20, 2012), at pp. 1-8, 2-141, 2-142 ("Based on IP2 and IP3's location within the State's Coastal Zone, license renewal of IP2 and IP3 will require a State coastal consistency certification").

001-5 Under Title 10 of the Code of Federal Regulations (10 CFR) 51.20(b)(2) and the National Environmental Policy Act of 1969, as amended (NEPA), the renewal of a power reactor operating license requires preparation of an environmental impact statement (EIS) or a supplement to an existing EIS. An EIS is prepared for any action determined to be a major Federal action significantly affecting the quality of the human environment. In general, an EIS contains detailed analyses of the reasonably foreseeable environmental impacts of the proposed action, and the environmental impacts of alternatives to the proposed action, and involves extensive public participation and coordination with local, State, and other Federal agencies. Whether or not a new water quality certification or a CZMA consistency determination is needed for the NRC to issue a renewed license is not within the scope of this FSEIS Supplement. See Section 1.5 of the FSEIS for further discussion of this issue.

001-5

12

Federal agency until the state or its designated agency [DOS] has concurred with the applicant's certification."[59]

001-5

Based on the forgoing, NRC Staff's revised Draft FSEIS Supplement contains flawed analyses and conclusions, and, as a result, NRC has yet to fully and adequately comply with NEPA in relation to the proposed license renewal of Indian Point.

Thank you for your consideration.

Respectfully submitted,

Deborah Brancato

Deborah Brancato
Staff Attorney

Phillip Musegaas, Esq.
Hudson River Program Director

---

[59] Federal regulations at 15 C.F.R. Part 930 sets forth these procedures; notably, a federal determination is no substitute for the State determination.

13

Comment response on preceding page(s).

UNITED STATES DEPARTMENT OF COMMERCE
National Oceanic and Atmospheric Administration
NATIONAL MARINE FISHERIES SERVICE
NORTHEAST REGION
55 Great Republic Drive
Gloucester, MA 01930-2276

AUG - 7 2012

Michael Wentzel
Projects Branch 2
Division of License Renewal
Office of Nuclear Reactor Regulation
U.S. Nuclear Regulatory Commission
Washington, D.C. 20555-0001

Dear Mr. Wentzel,

Your June 26, 2012, letter requests comments on the draft supplement to the final plant-specific Supplement 38 to NUREG-1437, "Generic Environmental Impact Statement for License Renewal of Nuclear Plants" (GEIS), regarding the license renewal of Indian Point Nuclear Generating Unit Nos. 2 and 3 (IP2 and IP3). This is a supplement to the FSEIS you published in December 2010.

We have reviewed the document and have no substantive comments. The description of the information that has become available since the FSEIS was published is consistent with our understanding of the available information. Also, the description of the Endangered Species Act section 7 consultation that was completed in 2011 and reinitiated in 2012 appears complete and accurate.

Thank you for the opportunity to comment on this document. We look forward to continuing to work with you on the ongoing consultation to consider effects of continued operations of IP2 and IP3 on shortnose and Atlantic sturgeon. Please contact Julie Crocker in our Protected Resources Division if you have any questions regarding this letter (978-282-8480 or Julie.Crocker@noaa.gov).

Sincerely,

John K. Bullard
Regional Administrator

002-1

Entergy

Entergy Nuclear Northeast
Indian Point Energy Center
450 Broadway, GSB
P.O. Box 249
Buchanan, NY 10511-0249
Tel (914) 254-2055

Fred Dacimo
Vice President
Operations License Renewal

NL-12-121

August 20, 2012

Chief, Rules, Announcements, and Directives Branch
Division of Administrative Services
Office of Administration
Mailstop: TWB-05-B01M
U.S. Nuclear Regulatory Commission
Washington, DC 20555-0001

Subject: Comments on Draft Supplement to the Final Plant-Specific Supplement 38 to
             NUREG-1437, Regarding License Renewal of Indian Point Nuclear Generating
             Unit Nos. 2 and 3
             Docket Nos. 50-247 and 50-286
             License Nos. DPR-26 and DPR-64

Reference: 1. June 2012 Draft Supplement 38 to the Generic Environmental Impact
                Statement for License Renewal of Nuclear Power Plants Regarding Indian
                Point Nuclear Generating Unit Nos. 2 and 3.

            2. March 18, 2009 Comments by Entergy regarding NUREG-1437, Draft
                Supplement 38. (ADAMS Accession No. ML091040133).

            3. March 29, 2011 Comments by Entergy regarding Final Supplemental
                Environmental Impact Statement Indian Point Nuclear Generating Unit Nos. 2
                & 3, Docket Nos. 50-247 and 50-286 License Nos. DPR-26 and DPR-64.
                (ADAMS Accession No. ML110980073).

Dear Sir or Madam:

Entergy Nuclear Indian Point 2, LLC, Entergy Nuclear Indian Point 3, LLC, and Entergy Nuclear
Operations, Inc. (collectively, "Entergy"), respectfully submits the following comments
("Comments") on certain portions of the June 2012 draft supplement (the "Draft Supplement") to
the December 2010 Final Supplemental Environmental Impact Statement ("FSEIS") prepared by
Nuclear Regulatory Commission ("NRC") Staff, and its consultants, for the Indian Point Nuclear
Generating Unit Nos. 2 and 3 ("Indian Point") License Renewal Application, assessing the
potential impacts of entrainment, impingement and thermal shock, including associated
mitigation (collectively, "Aquatic Issues"). These Comments relate to certain matters in the Draft
Supplement that NRC Staff may want to address before its finalization.

We commend NRC Staff for its work in the Draft Supplement, particularly including NRC Staff's:

(handwritten annotations:)

Received
8/28/12
3:00pm

7/6/2012

77 FR 40091
①

Sunsi Review Complete
Template= ADM-013

EFADS= ADM-03
Ell= m. wenzel (msu2)

003-1 Regarding Entergy's 2009 comments on the draft SEIS, the staff addressed those comments in the FSEIS and modified its entrainment and impingement analysis methods in the FSEIS in response to new information and comments submitted on the DSEIS. Regarding Entergy's 2011 comments on the FSEIS, the staff considered those comments in depth in preparing the FSEIS. This supplement to the FSEIS discusses only those comments for which new information could change any conclusions in the FSEIS.

The NRC's 2012 "Technical Analysis and Support for Generic Environmental Impact Statement for License Renewal of Nuclear Plants, Supplement 38, Regarding Indian Point Nuclear Generating Unit Nos. 2 and 3, Volume 4 Draft Report for Comment (NUREG-1437, Suppl. 38, Vol. 4, June 2012)" (ML12257A346) presents the NRC staff's assessment of Entergy's 2011 comments and the analyses the staff conducted to better understand the implications of the comments and how these comments might affect the staff's FSEIS conclusions. The staff did not find any new and significant information in Entergy's previous comments beyond the information that the staff has already addressed.

(1) corrections to impingement and entrainment data presented in the FSEIS and related conclusions; (2) revised conclusions regarding the absence of potential thermal shock and conclusions of SMALL thermal impacts as a function of Indian Point's compliance with New York State thermal water quality standards; and (3) update of the status of the NRC Staff's consultation with the National Marine Fisheries Service ("NMFS") under Section 7 of the Endangered Species Act ("ESA"), including NMFS's findings that continued operation of Indian Point would not adversely affect shortnose sturgeon. See Draft Supplement: pp. iii (summary); 3-17 (corrections of impingement and entrainment data, including to reflect assessment of SMALL potential impacts to spottail shiner); 17-21 (thermal impacts of Indian Point should be SMALL) and 23-26 (completion of NMFS consultation for shortnose sturgeon and update on re-initiation of consultation for recent listing of Atlantic sturgeon). Entergy concurs with these findings and conclusions in the Draft Supplement, except as noted below.

Specifically, Entergy requests that the NRC Staff revisit and address the comments that Entergy previously submitted, respectively dated March 18, 2009 and March 29, 2011, on the December 2008 Draft Supplemental Environmental Impact Statement ("DSEIS") and FSEIS, copies of which are expressly incorporated as if fully set forth here. These prior comments, including the technical appendices to those comments evaluating nearly four decades of biological monitoring of fish species in the Hudson River, establish that any potential impingement and entrainment impacts of Indian Point's continued operations during license renewal are properly considered SMALL to all identified fish species, including on Table 4-4. See Supplement, p. 9. To that end, we respectfully request that NRC Staff's findings of potential impingement and entrainment impacts in the Draft Supplement, including in Table 4-4, regarding alewife/blueback herring (evaluated as "river herring" at life stages susceptible to entrainment at Indian Point), hogchoker, rainbow smelt, weakfish and white perch, be determined to be SMALL.

003-1

We appreciate NRC Staff's efforts in this regard, and respectfully request that it implement these Comments when it publishes Final Supplement 38. There are no commitments identified in this submittal. Should you have any questions regarding these Comments, please contact Dara Gray at (914) 254-8414.

Sincerely,

FRD/mb

cc: Mr. William Dean, Regional Administrator, NRC Region I
Mr. Sherwin E. Turk, NRC Office of General Counsel, Special Counsel
Mr. Dave Wrona, NRC Branch Chief, Engineering Review Branch I
Mr. Robert F Kuntz, NRC Sr. Project Manager, Division of License Renewal
Mr. Douglas Pickett, Senior Project Manager, NRC NRR DORL
Mr. Michael Wentzel, NRC Environmental Project Manager, IPEC License Renewal
NRC Resident Inspector's Office
Ms. Bridget Frymire, NYS Dept. of Public Service
Mr. Francis J. Murray, Jr., President and CEO NYSERDA

004-1 This comment expresses support for the updates in the draft supplement to the FSEIS and for the renewal of the IP2 and IP3 operating licenses. This comment does not provide any new information; therefore, no changes were made to this supplement to the FSEIS in response to this comment.

RADReviewed
8/17/2012

7/06/2012
77 FR 40091
②

NRCREP Resource

From:	Arthur "Jerry" Kremer [info@area-alliance.org]
Sent:	Friday, August 17, 2012 2:44 PM
To:	NRCREP Resource
Subject:	Response from "Comment on NRC Documents"

Below is the result of your feedback form. It was submitted by

Arthur "Jerry" Kremer (info@area-alliance.org) on Friday, August 17, 2012 at 14:43:59

Document_Title: Generic Environmental Impact Statement for License Renewal of Nuclear Plants: Supplement 38 Regarding Indian Point Nuclear Generating Unit Nos. 2 and 3 - Draft Report for Comment (NUREG-1437, Supplement 38, Volume 4)

Comments: August 17, 2012

Chief, Rules, Announcements, and Directives Branch Division of Administrative Services Office of Administration Mail Stop: TWB-05-B01M U.S. Nuclear Regulatory Commission Washington, DC 20555-0001

Dear Sir or Madam:

I am writing on behalf of the New York Affordable Reliable Electricity Alliance (New York AREA) to express support for your revisions to the Generic Environmental Impact Statement for Indian Point Energy Center's Units 2 and 3, as outlined in NUREG- 1437, Supplement 38, Volume 4, draft supplement to final.

The revised supplement to the EIS indicates that the initial assessment of the Indian Point's impact to the Hudson River was greatly over estimated. Specifically, NRC staff overestimated the entrainment losses for each of the representative important species studied in the analysis by a factor of 1000.

NRC also revised conclusions regarding the impact of thermal discharge from Indian Point Units 2 and 3, based on information provided by the New York State Department of Environmental Conservation. The New York State DEC's findings indicate that the discharge from Indian Point Units 2 and 3 is in compliance with its water quality standards and criteria for thermal discharges.
As stated on page 20, lines 25-28, "NYSDEC and NYSDEC's (2011) conclusions regarding studies provide reasonable assurance that the IP2 and IP3 discharge is in compliance with NYSDEC's water quality standards and criteria for thermal discharges."

The report's revisions show that the slightly heated water released from Indian Point has a "small" impact to the Hudson River and confirms our belief that there is no environmental reason precluding the plant from having its operating licenses renewed.

Of particular note, the National Marine Fisheries Service finds that the shortnose sturgeon is not threatened. As stated from page 23, line 42, to page 24, line 4, "In its biological opinion, NMFS concluded that shortnose sturgeon are likely to avoid the small area of water elevated above the species preferred temperature range and that — it is extremely unlikely that these minor changes in behavior will preclude shortnose sturgeon from completing any essential behaviors such as resting, foraging or migrating or that the fitness of any individuals will be affected."

The license renewal of Indian Point is important for New York's environment, in particular our air quality. Indian Point's continued operation reduces New York's need for fossil fuels, thereby mitigating carbon and other toxic

004-1

SWISE Review Complete
Template=ADM-013

K-REDS=ADM-03
Code= M. Wentzel (NSIR)

1

A-19

Comment response on preceding page(s).

emissions. In light of New York's often poor air quality and continuing non-compliance with the federal Clean Air Act, it is imperative that the plant obtain license renewal. | 004-1 |

Thank you for your time.

Sincerely,

Arthur J. Kremer

Chairman
New York AREA

organization: New York Affordable Reliable Electricity Alliance (New York AREA)

address1: 114 West 47th Street

address2: 19th Floor

city: New York

state: NY

zip: 10036

country: United States

phone: 212-683-1203

2

005-1 This comment is similar to comment 001-2. The staff's assessment of aquatic species population trends in both the FSEIS and the supplement to the FSEIS uses empirical monitoring study data from the Hudson River Monitoring Program from 1974 through 2005 (the most recent data then available), and its assessment of thermal impacts in the supplement uses data from a study that Entergy submitted to the NYSDEC in 2011. The staff's analyses based on these empirical studies take into account any changes to the fish populations in the Hudson River that have occurred over the years. The impingement and entrainment data used in the strength of connection analyses (dated from 1975 through 1990) are the best and most recent site-specific data available. The staff has no reason to believe that the conditional impingement and entrainment mortality rate estimates based on those data would be different today, and the commenter presents no information to suggest that such rates have changed. The staff made no further changes to this supplement to the FSEIS in response to this comment.

Regarding the approach used in the staff's analysis, the staff used a weight of evidence approach adapted from EPA's guidelines for ecological risk assessment rather than a single species fisheries modeling approach. The staff acknowledges that it could have undertaken a broader, even more holistic analysis had data been available for aquatic non-fish populations. The Hudson River Monitoring Program on which the staff based its trend analyses is an extensive and well-designed monitoring program that samples many fish species; the staff found it provides reliable and substantial data that are adequate to support an impact assessment for the purposes of NEPA.

New York State Department of Environmental Conservation
Office of General Counsel, 14th Floor
625 Broadway, Albany, New York 12233-1500
Fax: (518) 402-9018
Website: www.dec.ny.gov

Joe Martens
Commissioner

August 20, 2012

Cindy Bladey
Chief, Rules, Announcements, and Directives Branch
Office of Administration
Mail Stop: TWB-05-B01M
U.S. Nuclear Regulatory Commission
Washington, D.C. 20555-0001

Re: Draft Supplement to Supplement 38 to the Generic Environmental Impact Statement for
License Renewal of Nuclear Plants, Regarding Indian Point Nuclear Generating Unit
Nos. 2 and 3, Draft Report for Comment dated June 2012 (NUREG-1437; Supplement
38, Vol. 4); 77 Fed. Reg. 40091 (July 6, 2012)

Dear Ms. Bladey:

On behalf of the New York State Department of Environmental Conservation
("NYSDEC"), please accept the following comments regarding the U. S. Nuclear Regulatory
Commission's ("NRC") *Draft Supplement to Supplement 38 to the Generic Environmental
Impact Statement for License Renewal of Nuclear Plants, Regarding Indian Point Nuclear
Generating Units Nos. 2 and 3, Draft Report for Comment dated June 2012* (NUREG-1437;
Supplement 38, Vol. 4) ("Draft Report"). NYSDEC appreciates the efforts of NRC Staff to
augment the record of the Final Supplemental Environmental Impact Statement so that it may
consider new data, analyses, and comments from various sources. Although NYSDEC Staff
concur with NRC Staff's conclusions in some respects, NYSDEC Staff's decision to continue its
reliance on out of date entrainment and impingement data, without requiring production of more
current data into the record and incorporating that into its analysis to make its findings with
respect to entrainment and impingement impacts is inconsistent with the National Environmental
Policy Act ("NEPA") and associated Council on Environmental Quality ("CEQ") and NRC
regulations and reflects a fundamental error in regulatory judgment that infects other aspects of
the NRC's relicensing review. Moreover, NRC Staff should expand the scope of its
environmental review and include a thorough analysis of the environmental impacts of severe
accidents at the Indian Point facilities on water resources and the alternatives to mitigate such
impacts.

NRC Staff's Draft Report corrects a mathematical error concerning entrainment and
impingement field data (see "Technical Review of FSEIS for Indian Point Nuclear Generating
Unit Nos. 2 and 3" [AKRF 2011b]). However, this correction overlooks the fact that the
foundational data base for entrainment and impingement at Indian Point Units 2 and 3 is more
than 25 years old and out of date (i.e., entrainment or impingement data have not been collected
at Indian Point since 1987), providing an inadequate basis for determining the gravity and harm
to be accorded to this adverse environmental impact under NEPA.

NRC Staff's expressed purpose in this stage of its NEPA review was to conduct an
analysis that is "more holistic than a general fisheries biology approach." However, the data on
which NRC Staff appears to rely and the approach that NRC Staff undertook in the June 2012
Draft Report was precisely that: a general fisheries approach, relying heavily on the Hudson

005-1

A-21

Comment response on preceding page(s).

2.

River Monitoring Program data. *See* Final Supplemental EIS, at H-13. The underlying analytical error committed by NRC Staff during this review process was that it failed to obtain appropriate and timely data to accomplish this stated purpose. Thus, NRC's NEPA analysis has necessarily failed to accomplish that goal.

A full and complete NEPA analysis requires that a thorough and temporally relevant study be conducted to collect data for determining the potential impacts of entrainment and impingement over the proposed relicensing term. NRC regulations contemplate preparing a supplement to a final environmental impact statement when, in NRC Staff's opinion, preparation of a supplement will further the purposes of NEPA. 50 C.F.R. § 51.92(b); 40 C.F.R. §§1500.1(b), 1502.22, 1502.24. (CEQ regulation provide that, when an agency is conducting an EIS, it must make clear where there are gaps in relevant information due to incomplete or unavailable information and take steps to remedy such gaps or, relevant information cannot be obtained, identify its relevance to evaluating the reasonably foreseeable significant adverse impacts on the human environment [40 C.F.R. §1502.22], and must insure the "professional integrity, including scientific integrity, of the discussions and analyses in environmental impact statements." 40 C.F.R. §1502.24). *See also* the Draft Report, Executive Summary, p. ix.

The NRC is thus obligated to ensure the timeliness, currency and quality of data supporting its NEPA analyses. However, NRC Staff have not required Entergy to perform the aforesaid data gathering. NYSDEC brought this fundamental failure to expand the data base to the NRC's attention in its May 26, 2011 comments on the Final Supplemental EIS for the License Renewal for Indian Point Unit Nos. 2 and 3 on pages 18 and 19. Current assessment is critical. The Hudson River's fish community and habitat has changed significantly since the 1980s, and the NRC's failure to require that the applicant produce data reflecting those changes respectively renders flaws the integrity of any entrainment and impingement impact analysis with

005-1

In the NEPA context, courts have consistently rejected the use of data more than ten years old, that measures a vital aspect of the cumulative environmental effects of past and current usage, or affects a decision resulting in a serious environmental impact. *See, e.g. Lands Council v. Forester of Region One of the United States Forest Serv.*, 395 F.3d 1019, 1031 (9th Cir. 2005) (holding that the use of 13-year-old trout habitat information prevented an accurate impact assessment of the project); *Sierra Club v. U.S. Dep't of Agriculture*, 1995 U.S. Dist. LEXIS 21507, *39 (S.D. Ill. Sept. 25, 1995) (rejecting the use of 10-year-old songbird population data when more recent data should have been gathered).

Courts have also been hesitant to approve decisions based on old data that is vital in determining the cumulative effect of past and current land use. For example, the U.S. Court of Appeals for the 9th Circuit has confirmed the need for a supplemental EIS where new scientific evidence about the impact of logging on the survival of the Northern Spotted Owl was available. *See, Portland Audubon Society v. Babbitt*, 998 F.2d 705 (9th Cir. 1993). Lacking that information, including recent scientific developments, "[t]he existing Timber Management Plans [did] not adequately address the impact of the individual planned timber sales on the survival of the northern spotted owl subspecies." *Id.* at 709. In *Lands Council*, the court cited the importance of evidence of current conditions in determining the cumulative effect of past and current timber harvesting on trout habitat and population, finding "the data about the habitat of the Westslope Cutthroat Trout was too outdated to carry the weight assigned to it." *Lands Council*, 395 F.3d at 1031. Similarly, in *Sierra Club* the court took issue with the use of outdated songbird population data, which was used as an indicator to monitor population trends in general. *Sierra Club*, LEXIS at *39. Also, the U.S. District Court for the District of Washington rejected data precisely because it pre-dated the effects on at-risk species of

005-2 As noted in the comment, a massive earthquake off the east coast of Honshu, Japan, produced a devastating tsunami that struck Fukushima. In response to the earthquake, tsunami, and resulting reactor accidents at Fukushima Dai-ichi (hereafter referred to as the "Fukushima events"), the Commission directed the staff to convene an agency task force of senior leaders and experts to conduct a methodical and systematic review of relevant NRC regulatory requirements, programs, and processes. Based on the agency's current knowledge of the Fukushima events, they do not provide a seriously different picture of the environmental impacts of severe accidents (as compared to the severe accident parameters analyzed in the GEIS (e.g., GEIS Chapter 5)), so as to require specific consideration in this FSEIS supplement. Nevertheless, the NRC will continue to evaluate the need to make improvements to existing regulatory requirements based on the task force report and additional studies and analyses of the Fukushima events as more information is learned. To the extent that any revisions are made to NRC regulatory requirements, they would be made applicable to nuclear power reactors generally, regardless of whether or not they have a renewed license. Therefore, no additional analyses have been performed in this FSEIS Supplement as a result of the Fukushima events.

This comment provided no new or significant information regarding the information or analysis in this supplement to the FSEIS that would challenge the conclusions of the supplement; therefore, no change was made to this supplement to the FSEIS.

Fax Server   8/20/2012 3:07:15 PM   PAGE   4/004   Fax Server

3.

disturbances caused by timber harvesting, and required the agencies involved to conduct full, accurate species reviews on all species protected by the "Survey and Manage" program before it could be eliminated. *Conservation Northwest v. Rey*, 674 F. Supp. 2d 1232, 1252-1253 (D. Wash. 2009).

Here, the 1987 data relied on by NRC Staff concerning the environmental impacts caused by Indian Points Unit 2 and Unit 3 diversion of water is obviously stale and universally recognized by regulators to be out of date, yet NRC Staff has not informed its own environmental analysis and decision making by requiring the submission of current, more relevant data. Without serious consideration of data representing the present impacts associated with entrainment and impingement from Indian Point Unit 2 and Unit 3, the NRC's final NEPA analysis and NEPA recommendation regarding Entergy's application for license extension will not be supported by sufficient and more accurate scientific information. [005-1]

Accordingly, the NYSDEC therefore respectfully requests that NRC forego finalization of the Draft Report and direct Entergy to obtain comprehensive and timely data and to allow a meaningful assessment of the current nature and extent of adverse impacts from Indian Point Unit 2 and Unit 3 associated with entrainment and impingement.

Moreover, given the impacts caused by the recent multi-reactor severe accidents in Japan, the NYSDEC respectfully requests that NRC review the potential environmental impacts of severe accidents involving the Indian Point reactors and spent fuel pools on surface waters and drinking water resources in the 50 miles surrounding the Indian Point site and explore alternatives to mitigate such impacts. [005-2]

Respectfully submitted,

Edward F. McTiernan
Deputy Counsel

cc: IndianPoint.EIS@NRC.gov

006-1 The staff acknowledges that new impingement and entrainment data could increase the certainty of the staff's conclusions. In numerous places in the FSEIS, the staff pointed out the lack of confirmatory studies on impingement mortality. The New York State Department of Environmental Protection (NYSDEC), not the NRC, regulates construction and operation of cooling water intake structures under the Clean Water Act and both Federal and New York State regulations. Therefore, the NYSDEC can require further studies on impingement and entrainment mortality rates, should it choose to do so. Although additional impingement mortality studies might increase the certainty of the staff's findings, the staff has no reason to believe that the conditional impingement and entrainment mortality rate estimates that it used would be significantly different today, and the commenter presents no information to suggest that such rates would have changed. In addition to these data, the staff utilized data for the period of 1974 through 2005 from the Hudson River Monitoring Program, which is an extensive and well-designed monitoring program that samples many fish species, affording the staff a high degree of confidence in its findings. The staff made no further changes to this supplement to the FSEIS in response to this comment.

UNITED STATES ENVIRONMENTAL PROTECTION AGENCY

REGION 2
290 BROADWAY
NEW YORK, NY 10007-1866

AUG 14 2012

Cindy Bladey
Chief, Rules, Announcements, and Directives Branch
Office of Administration
Mail Stop: TWB-05-B01M
U.S. Nuclear Regulatory Commission
Washington, D.C. 20555-0001

Rating: EC-2

Dear Ms. Bladey:

In accordance with Section 309 of the Clean Air Act and the National Environmental Policy Act (NEPA), the U.S. Environmental Protection Agency (EPA) has reviewed the Nuclear Regulatory Commission's (NRC) Draft Supplement to Supplement 38 to the Generic Environmental Impact Statement for License Renewal of Nuclear Plants (CEQ#20120215), regarding the renewal of operating licenses for an additional 20 years of operation for the Indian Point Nuclear Generating Units Nos. 2 and 3 (IP2 and IP3). IP2 and IP3 are located in Westchester County in the Village of Buchanan, New York, approximately 24 miles north of New York City.

Subsequent to NRC's release of the final supplemental environmental impact statement (FSEIS) for relicensing IP2 and IP3 in December 2010, NRC staff identified new information warranting changes to its assessment in the FSEIS. This included information on the entrainment and impingement field data units of measure, the completion of a study characterizing the IP2 and IP3 thermal plume, and completion of the Endangered Species Act (ESA) consultation on shortnose sturgeon (*Acipenser brevirostrum*).

EPA notes the changes made the appropriate units for the impingement and entrainment information in the FSEIS. While it does not significantly alter the conclusions made by the NRC staff in the FSEIS, and notwithstanding previous NRC responses to our comments, EPA still remains concerned about the aquatic impacts of cooling water intake and discharge at IP2 and IP 3. New impingement/entrainment data would have provided NRC and others with the information necessary to determine the level of significance of the impacts with more certainty. EPA has no comments on the thermal study or ESA consultation.

Received
8/28/12
3:00 pm

7/4/29/12
77 FR 44091   8

006-1

E-RIDS= ADM-03
Cee = m. Wentzel
(msur)

50VSI Review Complete
Template = ADM-013

Internet Address (URL) • http://www.epa.gov
Recyclable/Recyclable •Printed with Vegetable Oil Based Inks on Recycled Paper (Minimum 50% Postconsumer content)

We appreciate the opportunity to comment on the document. If you have any questions, please call Lingard Knutson of my staff at (212) 637-3747.

Sincerely,

Judy-Ann Mitchell, Chief
Sustainability and Multimedia Programs Branch
Clean Air and Sustainability Division

cc: Paul Giardina, CASD-RIAB

United States Department of the Interior

OFFICE OF THE SECRETARY
Office of Environmental Policy and Compliance
408 Atlantic Avenue – Room 142
Boston, Massachusetts 02110-3334

August 20, 2012

9043.1
ER 12/488

Cindy Bladey, Chief,
Rules, Announcements, and Directives Branch
Office of Administration
Mail Stop: TWB-05-B01M
U.S. Nuclear Regulatory Commission
Washington, DC 20555

RE: **COMMENTS**
Docket ID NRC-2008-0672
Draft Supplement to Supplement 38
Indian Point Nuclear Generating Units 2 and 3
Westchester County, New York

Dear Ms. Bladey:

The U.S. Department of the Interior (Department) has reviewed the Draft Supplement to Supplement 38 to the Generic Environmental Impact Statement for License Renewal of Nuclear Plants (GEIS), NUREG-1437, regarding the renewal of operating licenses DPR-26 and DPR-64 for an additional 20 years of operation for Indian Point Nuclear Generating Units 2 and 3, Buchanan, Westchester County NY (Docket ID NRC-2008-0672). The Department has no comment on the Draft Supplement.

Thank you for the opportunity to review and comment on this supplement. Please contact me at (617) 223-8565 if I can be of assistance.

Sincerely,

Andrew L. Raddant
Regional Environmental Officer

007-1

7/26/2012
77 FR 40091

(4)

ERIC T. SCHNEIDERMAN
ATTORNEY GENERAL

STATE OF NEW YORK
OFFICE OF THE ATTORNEY GENERAL

DIVISION OF SOCIAL JUSTICE
ENVIRONMENTAL PROTECTION BUREAU

August 20, 2012

Cindy Bladey
Chief, Rules, Announcements, and Directives Branch
Office of Administration
Mail Stop: TWB-05-B01M
U.S. Nuclear Regulatory Commission
Washington, D.C. 20555-0001

Re: Draft Supplement to Supplement 38 to the Generic Environmental Impact Statement
for License Renewal of Nuclear Plants, Regarding Indian Point Nuclear Generating Unit
Nos. 2 and 3, Draft Report for Comment dated June 26, 2012
NUREG-1437; Supplement 38, Vol. 4; 77 Fed. Reg. 40091 (July 6, 2012)
Docket Nos. 50-247-LR/50-286-LR

Dear Ms. Bladey:

Enclosed please find comments submitted by the State of New York Office of the
Attorney General concerning NRC Staff's draft supplemental environmental impact statement.

All citations and references mentioned in the comments are hereby incorporated by
reference. Should NRC Staff have difficulty obtaining any such citations and
references, they are requested to contact the Office of the Attorney General for the
State of New York for assistance. Please include these comments in the record for this
proceeding.

Respectfully submitted,

Signed (electronically) by

John J. Sipos
Assistant Attorney General
(518) 402-2251
john.sipos@ag.ny.gov

cc: IndianPoint.EIS@NRC.gov

THE CAPITOL, ALBANY, N.Y. 12224-0341 • PHONE (518) 473-3105 • FAX (518) 473- 2534 • WWW.AG.NY.GOV

A-27

COMMENTS BY THE NEW YORK STATE OFFICE OF THE ATTORNEY GENERAL

DRAFT SUPPLEMENT TO SUPPLEMENT 38 TO THE GENERIC ENVIRONMENTAL IMPACT STATEMENT

FOR LICENSE RENEWAL OF NUCLEAR PLANTS, REGARDING INDIAN POINT NUCLEAR GENERATING

UNIT NOS. 2 AND 3, DRAFT REPORT FOR COMMENT DATED JUNE 26, 2012

NUREG-1437; SUPPLEMENT 38, VOL. 4; 77 FED. REG. 40091 (JULY 6, 2012)

DOCKET NOS. 50-247-LR/50-286-LR

Submitted: August 20, 2012

008-1 As discussed in Section 5.1.2 of the FSEIS, the issue of severe accidents at nuclear power plants and the resulting impacts on the environment, including impacts to surface water and groundwater resources, was evaluated in NUREG-1437, Generic Environmental Impact Statement for License Renewal of Nuclear Plants (GEIS). The determination in the GEIS is that the probability-weighted consequences of severe accidents are SMALL for all plants; however, alternatives to mitigate severe accidents must be considered for all plants. As detailed in Section 5 and Appendix F of the FSEIS, the NRC evaluated Entergy's analysis of alternatives to mitigate severe accidents (referred to as SAMAs), and found that Entergy has appropriately identified areas in which risk could be reduced in a cost-beneficial manner.

This comment provided no new or significant information regarding the information or analysis in this supplement to the FSEIS that would challenge the conclusions of the supplement; therefore, no change was made to this supplement to the FSEIS.

Earlier this year, the State of New York wrote to NRC to express the State's concern about the narrow scope of NRC Staff's proposed supplemental review and revision of the Final Supplemental Environmental Impact Statement for the requested extension of the operating licenses for Indian Point Unit 2 and Indian Point Unit 3. See March 28, 2012 letter from J. Sipos to S. Turk (NRC) ML12090A609. Recently, NRC Staff released a draft report for public comment. NUREG-1437, Supplement 38, Vol. 4 (June 26, 2012) ML12174A244. In a notice in the Federal Register, Staff requested comments by August 20, 2012. 77 Fed. Reg. 40091 (July 6, 2012). The scope of the draft supplemental report remains far too narrow and is inconsistent with the National Environmental Policy Act and the implementing regulations promulgated by the Council on Environmental Quality and NRC.

NRC Staff's EIS Should Include a Site Specific Examination of the Impacts of Severe Accidents on Water Resources, Including Drinking Water Resources, Within 50 Miles of Indian Point and the Alternatives to Mitigate Such Impacts

NRC Staff prepared the draft report to examine information about environmental impacts of issuing renewed operated licenses to the Indian Point facilities on aquatic resources, but the supplemental review conducted by Staff examines only a narrow aspect of such impacts. The State previously requested that NRC Staff review the impacts of a severe accident on water resources, including drinking water resources. Such a review should include a thorough analysis of the value of such resources and the cost to decontaminate and remediate those resources, or provide replacement drinking water. Several reservoirs that provide drinking water resources for New York communities – including New York City – are located within 50 miles of the Indian Point facilities, yet NRC had not examined the environmental impacts that would result from a severe accident that deposited radionuclides into those water resources or the cost to

008-1

008-2 The issue of spent fuel storage accidents during the term of any renewed license was evaluated in the GEIS and determined to be of SMALL impact for all plants, and, thus, was designated a Category 1 issue for license renewal. The Commission reaffirmed this in 2008 upon denying two petitions for rulemaking seeking to challenge the Category 1 designation (73 FR 46204). The Commission's determination to consider SFP accidents on a generic basis in the GEIS rather then in a site-specific SEIS – was upheld in Massachusetts v NRC 522 F. 3rd 115, 127 (1st Cir. 2008). As the Commission stated in 2008, "given that the SFP risk level is less than that for a reactor accident, a SAMA that addresses SFP accidents would not be expected to have a significant impact on total risk for the site."

As a result of a recent decision by the U.S. Court of Appeals for the District of Columbia Circuit, in New York v. NRC, 681 F.3rd 471 (D.C. Cir., June 8, 2012), the issue of spent fuel storage, as it relates to the Commission's Waste Confidence Decision, (codified at 10 CFR 51.23), is subject to ongoing consideration by the Commission as a generic issue; this issue is outside the scope of this FSEIS suplement, and therefore, no change was made to address the issue in this supplement to the FSEIS.

008-1

008-2

decontaminate or replace those resources. Nor has NRC Staff examined alternatives to mitigate such impacts. Thus, NRC Staff has failed to conduct an adequate site specific analysis of the environmental impacts of a severe accident at Indian Point on water resources.

NRC Staff's EIS Should Examine the Impact of Severe Accidents to the Spent Fuel Pools at Indian Point and the Alternatives to Mitigate Such Impacts

Substantial amounts of spent nuclear fuel are contained in the spent fuel pools at Indian Point Unit 2 and Indian Point Unit 3. Recently, Entergy representatives disclosed that Entergy intendeds to maintain current dense storage practices at Indian Point during any relicensing term granted by NRC. Specifically, during the May 8, 2012 site visit to the Indian Point facilities by the Atomic Safety and Licensing Board, Entergy representatives made the following statements about Entergy's plans for spent nuclear fuel at Indian Point:

(A) All of the spent fuel generated during since the start of commercial operation of Indian Point Unit 3 remains in the Indian Point Unit 3 spent fuel pool (as of the date of the site visit) (stated differently, the Unit 3 spent fuel pool holds 36 years worth of spent nuclear fuel);

(B) Entergy has no current plans to construct an additional dry cask storage area (in addition to the existing dry cask storage area); and

(C) At the end of operation under any 20-year extension of the current operating licenses, Entergy estimates that the existing dry cask storage area would be filled to capacity and that the Indian Point Unit 2 spent fuel pool and the Indian Point Unit 3 spent fuel pool would be filled to capacity as well.

NRC Staff has not examined alternatives to the continued dense storage of spent nuclear fuel in the Indian Point spent fuel pools.

NRC can no longer maintain that dry cask storage and densely packed spent fuel pools provide comparable long term storage options. The differences between the two storage methods are illustrated by the following chart reflects NRC's list of priorities five days into the

008-2 The NRC continues to evaluate relevant regulatory requirements, programs, and processes in light of the reactor accidents at the Fukushima Dai-ichi nuclear power plant. Based on the agency's current knowledge of the Fukushima events, they do not constitute information that would reveal a seriously different picture of the environmental impacts of severe accidents (as compared to the severe accident parameters analyzed in the GEIS) so as to require specific consideration in this supplement to the FSEIS. The NRC will continue to evaluate the need to make improvements to existing regulatory requirements as more information is learned. To the extent that any revisions are made to NRC regulatory requirements, they would be made applicable to nuclear power reactors generally, regardless of whether or not they have a renewed license.

This comment provided no new or significant information regarding the information or analysis in this supplement to the FSEIS that would challenge the conclusions of the supplement; therefore, no change was made to this supplement to the FSEIS.

Fukushima disaster. The conditions at Fukushima Daiichi Unit 1, Unit 2, and Unit 3 were worse than that reached at any time during the Three Mile Island Unit 2 reactor accident, yet the NRC's higher priority was the Daiichi Unit 4 spent fuel pool.

Fukushima Daiichi Summary Display

Priority	Unit	STATUS AS OF 06:00 EDT (19:00 Local) - 03/16/2011
4	1	**Core Status** - Severe core damage (based on the amount of hydrogen generated). Radiation has been released. Possible RCS breach. (GE) Sea water injection to RPV. **Containment** - Primary apparently intact. Secondary Containment destroyed. **Spent Fuel Pool** - No information on SFP status.
3	2	**Core Status** – Severe core damage likely. Radiation release has occurred. Possible RCS breach (GE). Sea water injection to RPV. **Containment** - Primary apparently intact. Secondary Containment lost. **Spent Fuel Pool** – No information on SFP status. Some reports attribute smoke/steam coming from the SFP
2	3	**Core Status** - Severe core damaged (based on the amount of hydrogen generated). Radiation has been released. Possible RCS breach. (GE). Sea water injection to RPV. **Containment** - Primary apparently intact. Secondary Containment destroyed. **Spent Fuel Pool** – May be in the same condition as Unit 4 SFP below. (Monninger)
1	4	Core off-loaded to Spent Fuel Pool. Secondary Containment destroyed. Walls of SFP have collapsed. No SFP cooling is possible at this time. TEPCO requests recommendations. (Monninger)
5	5	Shutdown since January 3, 2011. Core loaded in RPV. RPV/SFP levels lower than normal and decreasing. Unit 6 D/G providing make-up water to Unit 5. (IAEA).
6	6	Shutdown since August 14, 2010. Core loaded in RPV. RPV/SFP levels lower than normal. Unit 6 D/G providing make-up water to Unit 5. (IAEA).

Source: NRC ADAMS Accession No. ML12080A196 (frame 259 of 782) (placed on public ADAMS on March 23, 2012) (highlight added). Moreover, there were concerns about radiation releases from the fuel rods in the Unit 4 spent fuel pool. *Id.* (frame 252 of 782 ("Rad release – possible from SFP")). Not only was the Daiichi Unit 4 spent fuel pool a top NRC priority, but it appears that the Daiichi Unit 4 spent fuel pool was not as densely packed as the two spent fuel pools at Indian Point Unit 2 and Indian Point Unit 3. In contrast, the spent nuclear fuel stored in the dry cask storage facility at Fukushima Daiichi was not included in NRC's list of priorities.

008-3 As part of the Federal government's National Response Framework, NRC is the Coordinating Agency for radiological events occurring at NRC-licensed facilities and for radioactive materials either licensed by NRC or under NRC's Agreement States Program. As the Coordinating Agency, NRC has technical leadership for the Federal government's response to the event. If the severity of an event rises to the level of General Emergency, or is terrorist-related, the Department of Homeland Security will take on the role of coordinating the overall Federal response to the event, while NRC would retain a technical leadership role; other Federal agencies who may respond to an event at an NRC-licensed facility, or involving NRC-licensed material, include the Federal Emergency Management Agency, the Department of Energy, the Environment Protection Agency, the Department of Agriculture, the Department of Health and Human Services, the National Oceanographic and Atmospheric Administration, and the Department of State.

Costs associated with nuclear incidents are governed by the Price-Anderson Nuclear Industries Indemnity Act (Price-Anderson Act; 42 U.S.C. 2210). The Price-Anderson Act is a Federal law that governs liability-related issues for non-military nuclear facilities in the United States. The main purpose of the Act is to provide prompt and orderly compensation to the public who may incur damages from a nuclear incident, no matter who might be liable.

008-3

NRC Staff's EIS Should Examine How the Federal Government and its Agencies Will Respond to Severe Accidents at Indian Point and Pay for the Decontamination of the New York Metropolitan Area Including its Water Resources

The State previously raised concerns about the federal government's ability to respond to a severe accident at Indian Point. Although Entergy has asked NRC to issue two operating licenses for the Indian Point reactors, spent fuel pools, and attendant facilities, it not clear that NRC has the desire, capability, or financial resources to respond to a severe accident at Indian Point and ensure the thorough decontamination of the New York metropolitan area including, but not limited to, its water resources – and drinking water resources – in the wake of such an accident.

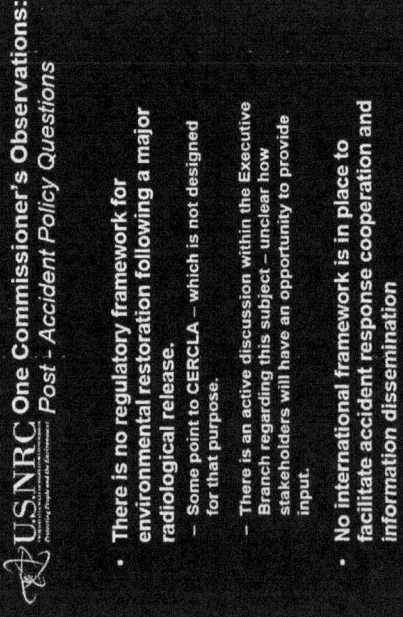

Power Point Presentation, Health Physics Society (Feb. 6, 2012).

A-32

008-3 cont'd Power reactor licensees are required to have the maximum level of primary insurance available from private sources (currently $375 million per 10 CFR 140.11) and are also required to participate in a Secondary Financial Protection Program. Under this program, should an accident at any participating power reactor result in injury or damage in excess of the maximum level of primary insurance, all power reactor operators will be charged a retrospective premium, up to a maximum of $111,900,000 per reactor per incident per 10 CFR 140.11. These insurance levels are subject to adjustments due to inflation at 5-year intervals. The last adjustment was made in August 2009.

There currently are 104 power reactors that participate in the Secondary Financial Protection program, creating a combined level of protection under both the primary and secondary layers of approximately $12 billion.

In the event of a nuclear incident involving damages in excess of the limits established in the Act, Price Anderson includes a provision that obligates Congress to take appropriate action to provide compensation for public liability claims.

This comment is outside the scope of the license renewal environmental review, and provided no new or significant information that would warrant a change to this supplement to the FSEIS.

008-4 This is a conclusion or summary of comments 008-1 through 008-3, discussed above.

This comment provided no new or significant information that would warrant a change to this supplement to the FSEIS.

Conclusion

NRC Staff's supplemental review of the environmental impacts of the issuing 20 year operating licenses to Indian Point Unit 2 and Unit 3 is far too narrow and inconsistent with the National Environmental Policy Act and regulations promulgated by the Council on Environmental Quality and NRC. NRC should reopen its NEPA review of the requested operating licenses and conduct a comprehensive site specific analysis of the environmental impacts caused by severe accidents at Indian Point as well as the means and resources to mitigate such impacts.

008-4

John Sipos
Assistant Attorney General
State of New York

A.2   References

10 CFR 51. *Code of Federal Regulations*, Title 10, *Energy*, Part 51, "Environmental protection regulations for domestic licensing and related regulatory functions."

50 CFR 402. *Code of Federal Regulations*, Title 50, *Wildlife and Fisheries*, Part 402, "Interagency cooperation—Endangered Species Act of 1973, as amended."

73 FR 46204. U.S. Nuclear Regulatory Commission. The Attorney General of Commonwealth of Massachusetts, The Attorney General of California; Denial of Petitions for Rulemaking. *Federal Register* 73(154):46204-46213.

[NRC] U.S. Nuclear Regulatory Commission. 1996. Generic Environmental Impact Statement for License Renewal of Nuclear Power Plants. Washington, DC: NRC. NUREG–1437. May 1996. ADAMS Nos. ML040690705 and ML040690738.

[NRC] U.S. Nuclear Regulatory Commission. 2010. Generic Environmental Impact Statement for License Renewal of Nuclear Plants: Supplement 38, Regarding Indian Point Nuclear Generating Unit Nos. 2 and 3. Washington, DC: NRC. NUREG–1437, Supp. 38. December 2010. ADAMS Accession No. ML103270072.

[NRC] U.S. Nuclear Regulatory Commission. 2012. Technical Analysis and Support for Generic Environmental Impact Statement for License Renewal of Nuclear Plants, Supplement 38, Regarding Indian Point Nuclear Generating Unit Nos. 2 and 3, Volume 4 Draft Report for Comment (NUREG-1437, Suppl. 38, Vol. 4) June 2012. 98 pp. ML12257A346.

Price-Anderson Nuclear Industries Indemnity Act of 1957, as amended. 42 U.S.C §2210.

U.S. Court of Appeals for the D.C. Circuit. New York v. NRC, 681F.3rd471. June 8, 2012. Decision on Petitions for Review of Orders of the Nuclear Regulatory Commission.

NRC FORM 335
(12-2010)
NRCMD 3.7

U.S. NUCLEAR REGULATORY COMMISSION

BIBLIOGRAPHIC DATA SHEET

(See instructions on the reverse)

1. REPORT NUMBER
(Assigned by NRC, Add Vol., Supp., Rev., and Addendum Numbers, if any.)

NUREG-1437
Supplement 38
Vol. 4

2. TITLE AND SUBTITLE

Generic Environmental Impact Statement for License Renewal of Nuclear Plants, Supplement 38, Regarding Indian Point Nuclear Generating Unit Nos. 2 and 3, Final Report Supplemental Report and Comment Responses

3. DATE REPORT PUBLISHED

MONTH	YEAR
June	2013

4. FIN OR GRANT NUMBER

5. AUTHOR(S)

See Chapter 6

6. TYPE OF REPORT

Technical

7. PERIOD COVERED (Inclusive Dates)

8. PERFORMING ORGANIZATION - NAME AND ADDRESS (If NRC, provide Division, Office or Region, U. S. Nuclear Regulatory Commission, and mailing address; if contractor, provide name and mailing address.)

Division of License Renewal
Office of Nuclear Reactor Regulation
U.S. Nuclear Regulatory Commission
Washington, DC 20555-0001

9. SPONSORING ORGANIZATION - NAME AND ADDRESS (If NRC, type "Same as above", if contractor, provide NRC Division, Office or Region, U. S. Nuclear Regulatory Commission, and mailing address.)

Same as above

10. SUPPLEMENTARY NOTES
Docket Nos. 50-247 and 50-286

11. ABSTRACT (200 words or less)

This supplement to the final supplemental environmental impact statement (FSEIS) for the proposed license renewal of Indian Point Nuclear Generating Unit Nos. 2 and 3 incorporates new information that the U.S. Nuclear Regulatory Commission (NRC) staff has obtained since the publication of the FSEIS in December 2010.

This supplement includes corrections to impingement and entrainment data presented in the FSEIS, revised conclusions regarding thermal impacts based on newly available thermal plume studies, and an update of the status of the NRC's consultation under Section 7 of the Endangered Species Act with the National Marine Fisheries Service regarding the shortnose sturgeon (Acipenser brevirostrum) and Atlantic sturgeon (Acipenser oxyrinchus oxyrinchus).

12. KEY WORDS/DESCRIPTORS (List words or phrases that will assist researchers in locating the report.)

Indian Point Energy Center
Indian Point
Indian Point 3
Supplement to the Generic Environmental Impact Statement
SEIS
GEIS
National Environmental Impact Statement
NEPA
License Renewal

NUREG-1437, Supplement 38
Indian Point 2
Entergy Nuclear Operations, INC

13. AVAILABILITY STATEMENT
unlimited

14. SECURITY CLASSIFICATION

(This Page)
unclassified

(This Report)
unclassified

15. NUMBER OF PAGES

16. PRICE

NUREG-1437, Vol. 4
Supplement 38
Final Supplement

Generic Environmental Impact Statement for License Renewal of Nuclear Plants
Regarding Indian Point Nuclear Generating Unit Nos. 2 and 3

June 2013

www.ingramcontent.com/pod-product-compliance
Lightning Source LLC
Chambersburg PA
CBHW081830170526
45167CB00007B/2778